PETERSEN'S BOOK OF MAN IN SPACE
A GIANT LEAP FOR MANKIND

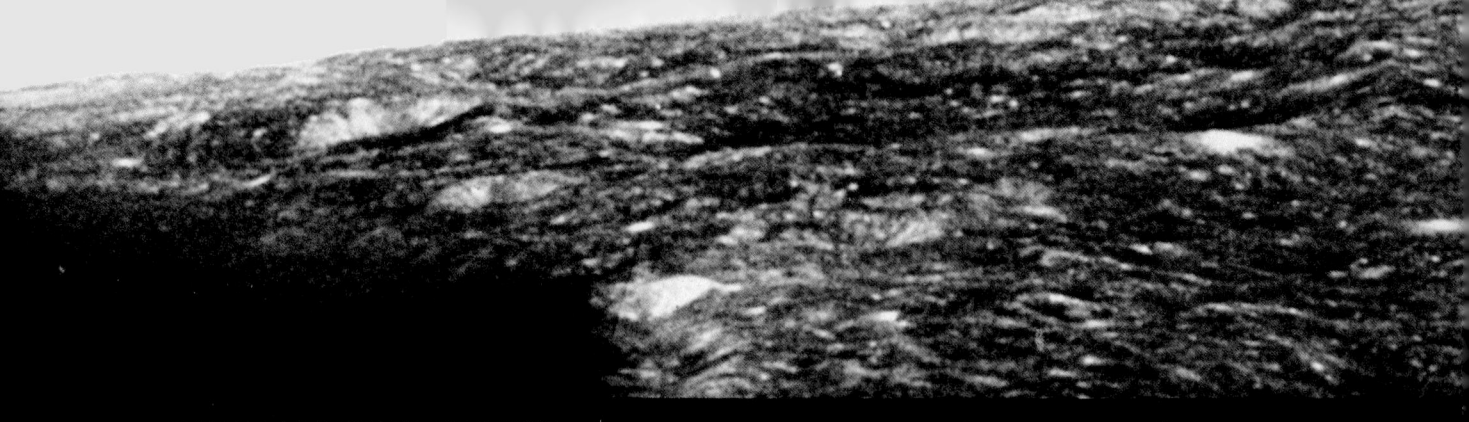

MAN IN SPACE Vol. 4 A GIANT LEAP FOR MANKIND

Edited by Al Hall and the General Subjects editors of Specialty Publications Division. Copyright ©1974 by Petersen Publishing Co., 8490 Sunset Blvd., Los Angeles, Calif. 90069. Phone: (213) 657-5100. All rights reserved. No part of this book may be reproduced without written permission. NASA photographs are government publications—not subject to copyright. Printed in U.S.A.

ACKNOWLEDGEMENTS

The editors of Petersen's MAN IN SPACE are indebted to the following agencies, companies, and individuals for their enthusiastic cooperation in helping locate and make available the numerous photos and hundreds of bits of information which add immeasurably to the scope and content of this volume. Our profuse thanks to: the National Aeronautics and Space Administration (which supplied all photos unless otherwise credited), and especially Les Gaver, Margaret Ware, and the Audio-Visual Branch, NASA Headquarters; Jack Stewart, Karla Scott, Donna Green, and NASA's Ames Research Center; John MacLeish, Jack Riley, Jim McBarron, and NASA's Johnson Space Center; Barbara Rogers, Dorothy Marine, and NASA's Kennedy Space Center; Anthony J. Longo and the Space Division of Rockwell International Corporation; Richard C. Dunne and the Grumman Corporation; Richard Coe and the International Latex Corporation; Walter Braun, Elsie Behmer, and Paillard, Inc.; Mike Gentry and Technicolor, Inc.; David M. Long and the Hamilton Standard Division of United Aircraft Corporation; and the Whirlpool Corporation.

COVER

Astronaut Edwin E. Aldrin Jr. poses beside the Lunar Module *Eagle* and the Solar Wind Experiment at Tranquility Base on man's first expedition to the moon. Photo by astronaut Neil Armstrong. Cover design by Pat Taketa.

INSIDE FRONT COVER

Nations around the world commemorate the efforts of the United States and the Soviet Union in space exploration with postage stamps. (From the collection of Kurt Raymond Hall.)

Library of Congress Catalog Card Number 74-81004 ISBN 0-8227-0075-1

Editor
AL HALL

Art Director
ROBERT I. YOUNG

Managing Editor
RICK BUSENKELL

Technical Editor
ALLEN BISHOP

Technical Editor
TERRY PARSONS

Technical Editor
JIM NORRIS

Design Artists
PAT TAKETA
DICK FISCHER
GEORGE FUKUDA

Contributing Editor
JILL FAIRCHILD

Contributing Editor
TERRIN MARDER

SPECIALTY PUBLICATIONS DIVISION

Hans Tanner/Editorial Director
Erwin M. Rosen/Executive Editor
George E. Shultz/Editorial Coordinator
Angie Ullrich/Secretary, Adminstrative
Holly Bjorseth/Secretary, Editorial

GENERAL SUBJECTS

Al Hall/Editor
Richard L. Busenkell/Managing Editor
Ronda Brown/Managing Editor
Harris R. Bierman/Associate Editor
Allen Bishop/Associate Editor
John T. Jo/Associate Editor
Terry Parsons/Associate Editor

AUTOMOTIVE

Spencer Murray/Editor
Judy Shane/Managing Editor
Jay Storer/Feature Editor
Jon Jay/Technical Editor
Tom Senter/Associate Editor

SPECIAL PROJECTS

Don Whitt/Editor
Steve Amos/Art Director
Ann C. Tidwell/Managing Editor
Jo Anne Peterson/Editorial Assistant

ART SERVICES

Robert I. Young/Art Director
Steve Hirsch/Artist, Design
Pat Taketa/Artist, Design
Dick Fischer/Artist, Design
George Fukuda/Artist, Design
Celeste Swayne-Courtney/Artist, Design
Nancy Quinn/Artist, Design
Ellen Clark/Artist, Design
Kathy Philpott/Artist, Design
Lloyd Haynes/Artist, Design

PETERSEN PUBLISHING COMPANY

R. E. Petersen/Chairman of the Board
F. R. Waingrow/President
Robert E. Brown/Sr. V.P., Corporate Sales
Herb Metcalf/V.P., Circulation Director
Dick Day/V.P., Automotive Publications
Phillip E. Trimbach/Controller-Treasurer
Robert Andersen/Director, Manufacturing
Al Isaacs/Director, Graphics
Bob D'Olivo/Director, Photography
Spencer Nilson/Director, Administrative Services
Larry Kent/Director, Corporate Merchandising
Ronald D. Salk/Director, Public Relations
William Porter/Director, Single Copy Sales
Jack Thompson/Director, Subscription Sales
Alan C. Hahn/Director, Market Development
Ralph L. Holt/Director, Editorial Research
Maria Cox/Manager, Data Processing Services
Robert Horton/Manager, Traffic
Harold L. Davis/Manager, Advertising Production
James J. Krenek/Manager, Purchasing
Mike Weldon/Manager, Editorial Production

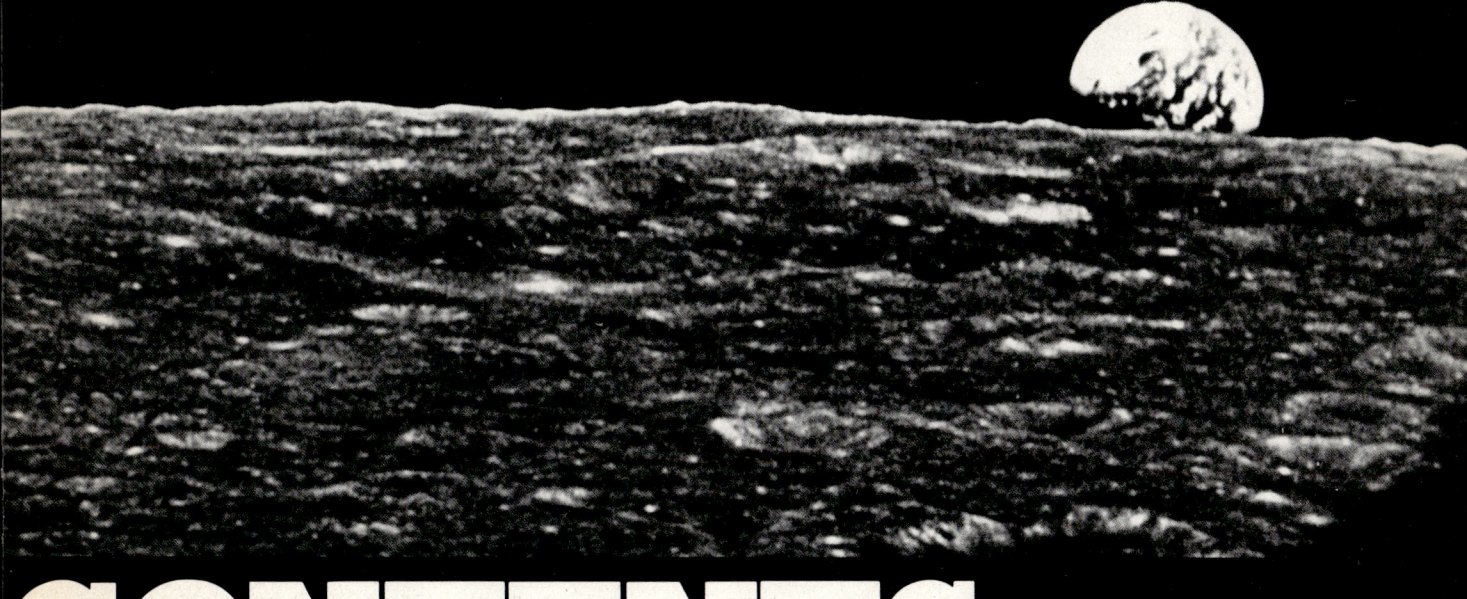

CONTENTS

4 INTRODUCTION

6 THE TRACKING NETWORK / How NASA keeps tabs on satellites and spacecraft

12 THE COMMAND & SERVICE MODULE / A guided tour of the Apollo crewmen's home in space

22 APOLLO 7 / Schirra, Eisele and Cunningham put Apollo through its paces

34 THE RED, WHITE & BLUE / Breakfast, lunch and snacks—the story of space food

40 APOLLO 8 / Borman, Lovell and Anders give earthlings a Christmas present

58 THE LUNAR MODULE / An ugly duckling designed for another world

70 APOLLO 9 / McDivitt, Scott and Schweickart rendezvous and dock with LM

86 SPACE PHOTOGRAPHY / A new breed cameras for a new kind of tourist

96 APOLLO 10 / Stafford, Young and Cernan stage a dress rehearsal

112 THE LAUNCH WINDOW / How NASA hits a moving target from a moving platform

114 SUIT-UP FOR SPACE / A new coat of armor shields man against a hostile environment

124 APOLLO 11 / A dream of centuries becomes reality—man walks on the moon

INTRODUCTION

This volume, fourth in this MAN IN SPACE series, continues the chronology of America's "race for space" with a look at the era from October 1968 through July 1969: A mere nine months! But those nine months represent a giant leap into space and, in fact, a giant leap for mankind.

AMERICA-6, RUSSIA-0

The year 1967, which marked the 10th anniversary of Sputnik's launching and the start of the "space race," had been a very bad year for the space programs of both America and Russia. In February 1967, the score was America-6, Russia-0. Theodore C. Freeman had been the first astronaut to lose his life. His plane crashed during a training mission in October of 1964. Charles A. Bassett II and Elliot M. See, Jr., also perished on a training flight. They had been selected as the prime crew for the Gemini IX mission and were enroute to McDonnell for two weeks of flight simulation training, but in attempting to land at the fog-shrouded St. Louis Municipal Airport adjacent to the McDonnell Aircraft Corporation facility their T-38 jet trainer veered out of the runway approach pattern, ricochetted off the roof of McDonnell's Building 101 and crashed in flames. Ironically, Building 101 housed the Gemini IX spacecraft that they were to ride into space. That was February 28, 1966. But January 27, 1967 was the blackest day in America's space program. That was the day a fire in the Apollo/Saturn 204 spacecraft claimed the lives of astronauts Virgil I. Grissom, Edward H. White and Roger B. Chaffee. And that was the day that Project Apollo almost ended before it had really begun. The A/S 204 Review Board and a Senate investigating committee had posed that threat. The Apollo 204 Review Board filed a report that criticized NASA and its contractors which was then picked up by the House Subcommittee on NASA Oversight. The summary of their findings caused NASA and North American Aviation more than mere embarrassment with statements the likes of overconfidence, shoddiness, complacency, the capsule's a death-trap, and inferences to NASA's attempts to soft-pedal the problems arising from the arbitrary deadline established by the late President Kennedy. NAA was chastized for shipping out a component or components that in some areas were poorly finished and in others were unfinished. NASA was condemned for accepting anything less than a properly finished product. The prime cause centered about the pure oxygen atmosphere in the space capsule, but in spite of the high risk it offered it was retained because the use of a two gas (oxygen-nitrogen) atmosphere as used by the Russians since their first manned flight was considered technically too difficult to handle. It seemed, at that moment, as though the Russians were still ahead in the race.

RUSSIA WINS

The A/S 204 disaster had definitely handed the Russians an advantage in the race to the moon. Both NASA and NAA had some problems to iron out in order to get their acts together. Getting the Apollo project back in shape was just part of NASA's problem, they also had to get the space program back on schedule. It would be almost two years from the A/S 204 fire until Americans would again be launched into space and meanwhile there were the Russians.

RUSSIA LOSES

After the spectacular successes of the Vostok and Voskhod missions, the Soviet space program had slowed a bit. The death of Sergei Korolev, chief of Russia's space program, in January 1966 had no doubt been a setback, but the brilliant engineer of Russia's space conquests had built a good foundation with Vostok and Voskhod. There was no reason not to expect the new Soyuz space craft to follow the same pattern. The launch was flawless, the mission quite successful and the return to earth couldn't help but be another routine landing. But Soyuz One's landing on April 24, 1967 was not as expected. A parachute malfunction moments after re-entry caused the capsule to plummet through the atmosphere, smash into the earth and snuff out the life of cosmonaut Vladimir Komarov. Less than 90 days after America's darkest hour, Komarov's death by system malfunction had cancelled out the Soviet Union's advantage. It would be a year and a half before Russia ventured to send another man into space and he would not have his eyes on the moon.

For some unexplained reason, the Russians redirected their space efforts toward earth orbits and a giant space station. Arguments raged among the experts that the Russians simply couldn't come up with the technology or hardware necessary for a lunar landing.

A FEW FIRSTS

Perhaps the experts had forgotten the long string of "firsts" which the Russians had already compiled? After all, the first satellite, Sputnik, was a product of Soviet technology. The first terrestrial creature to orbit the earth and test the safety of space for man was a Russian dog. This even before America had successfully launched its first satellite. In fact a joke popular at the time (circa 1958) kiddingly stated that the Russians were next going to orbit a dozen cows. The punch line was that it would be: "The first herd shot round the world!" An offhanded compliment to Soviet technology? The first man in orbit was Russian; ditto for the first woman. The first man to "walk in space" was Russian. The Russians were also first to rendezvous (Vostok 3 and 4 with Nikolayev and Popovich), to orbit a three-man spacecraft (Voskhod 1), to photograph the moon's hidden face, to launch a deep space probe, to hit the moon (Luna-2), to orbit spacecraft with extensive maneuvering capability (Polyot 1), to soft-land and broadcast close-up photos of the moon (Luna-9), to automatically dock in space (Kosmos 186 with Kosmos 188), to orbit a satellite around moon and return it to earth (Zond-5).

All of these firsts, and more, before America had launched the first manned Apollo seem to dispute the experts' claim that Russia lacked the technology and hardware. Experts and their predictions always abound. A good case in point was the furor created by Luna 3's photos of the back side of the moon in October 1959. The photos were immediately denounced as fakes by American experts. The simple truth was that although the photos had been retouched, they were not fakes and Americans unwilling to admit the Russians could be so far ahead of the U.S. in technology seized straws. It took nearly seven years and Lunar Orbiter 1 to prove the authenticity of the Russian photos, by which time most Americans had forgotten about the "expert" opinions of the Soviet photographs.

POLICY CHANGE

In direct contrast to the Americans' policy of keeping everyone aware of what was happening from launch through recovery, the Soviet Union's space achievements were only disclosed after they had been concluded...successfully. This policy no doubt saved a great deal of embarrassment, but led the press to view every announcement issued by the Russians with suspicion. But the Russians were eventually forced to be more open regarding their space achievements simply because they were competing with America. In Russia's eyes perhaps America's openness was foolhardy. After all, why invite the world to watch you fail? It's a gamble, but success is even sweeter when everyone knows the odds you face. The Russians reluctantly realized this after a while and followed America's lead...after a fashion.

THE LEADER'S LOST

However, the question is why did the Russians fade in the stretch run of the race to the moon. The inescapable conclusion is that Sergei Korolev's death marked the beginning of the end. Academician Sergei Pavlovich Korolev began working with rockets in 1931 as director of an enthusiast rocket club in Moscow known as GIRD. A year later he became director of all the GIRD chapters in the Soviet Union. Korolev's accomplishments included launching club built liquid-fueled rockets, a switch to designing rocket-powered aircraft during World War II, and back to peaceful rocket experiments after the war. He was, thus, eminently qualified to coordinate Russia's program to put man in space. As early as 1954, Korolev had headed a team of scientists and technicians who successfully launched animal subjects to heights of over 60 miles and recovered them safely. Coordinate, he did! It's not inconceivable that he faced the same problems that constantly beset the National Aeronautics and Space Administration regarding budgets, schedules, et al. And it doesn't require much imagination to picture Korolev supervising assembly and test of launch vehicles, spacecraft, training of cosmonauts, etc. Not that he was a one-man NASA, but from reading what little material is available on Soviet space efforts, it seems as though Korolev exercised far more supervision and control overall than his American counterpart. Little wonder, then, that with his death the Russians' lead in the space race dwindled.

COLUMBUS DIES

But the race paled into insignificance with the announcement that Yuri Gagarin was dead. The ever-smiling Columbus of the Cosmos became yet another victim of a training flight accident on March 27, 1968. Whether you were American, European, African or Russian, the news was shocking and the grief genuine. To those who knew him and worked with him, the grief must have been overwhelming. Most of the world only knew of him through pictures, but, as the old saying goes: "One picture is worth 1000 words." Some of the words conjured up by a picture of Yuri Gagarin are: warmth, happiness, honesty, modesty, courage and love of life. Life ended for Yuri and his co-pilot Vladimir Seregin, a qualified flight instructor, in the grinding crash of their MIG-15 jet. There is compensation in the fact that Gagarin's name has been written into history books for ages to come.

APOLLO AWAKENS

Others, however, were waiting impatiently for their chance to make history. In less than a year other names would join Gagarin's in history books. The American juggernaut was rolling. Man was on his way to the moon. The timid probing of Project Mercury helped pave the way for the bolder push of Project Gemini, but Project Apollo was to be the summation. There was still much to be done; equipment to be checked out and refined, techniques to rehearse and perfect, but America was rushing pell-mell toward the target selected by the late President John F. Kennedy. The commitment had been made and the goal was in sight. There was no way of knowing, at this time, that Russia was not still right in there. 1967 and 1968 had been a bad period for both America and Russia, but both nations were committed to the exploration of space and exploration has never come cheap. How many explorers had perished attempting to chart the wilderness? And how many sailors had lost their lives trying to conquer the oceans and seas? How many lives had it cost to master simple flight in the treacherous air above the earth?

OUR APPROACH

On the following pages you will find the chronology of the boldest exploration ever attempted and the steps which led up to it. "A GIANT LEAP FOR MANKIND" is written in present tense (like the other volumes in this series) as virtually a moment-by-moment account. There are several reasons for this approach, not the least of which is to give those who weren't able to experience these moments firsthand an idea of what it was like for those who did. Another reason is to give those who did experience them the benefit of an "instant replay."

It is our hope that in reading the following pages, those of you who thought you were familiar with the culmination of the space race will discover some new and fascinating aspects. And we hope that those of you who were not fortunate enough to follow it "live," will gain an understanding of the incredible technology it took and come away with a deep respect for America's ability to tackle the most complex task or seemingly insoluble problem and pursue it to a satisfactory conclusion.

Those who would decry this effort as frivolous waste might do well to reflect upon what Benjamin Franklin said in response to a pessimist at one of the first manned balloon ascensions in Paris almost 200 years ago who said loudly, "Of what possible use are these things." Franklin's reply was, "Of what possible use is a new-born babe?"

THE EDITORS

In order to keep a constant watch on our vehicles out in space, NASA has developed a worldwide network of tracking stations, the most sophisticated ever devised. This tracking system would have been the envy of navigators in the past ages. A global map (**1**) shows the network layout as arranged by NASA. Stations at these world-encircling points permit 24-hour contact with our spacecraft, both orbital and deep space types. The center of this chain is the Goddard Space Flight Center in Greenbelt, Maryland. Located in Goldstone, California is a 64-meter (210-foot) tracking antenna (**2**) capable of assimilating 16,200 bits (separate characters) of telemetry data per second, a capability eight times greater than that of the more common 85-foot antennas. A familiar sight on the high seas near launching and splashdown sites is a Russian trawler (**3**), most probably on tracking duty, as are the communications technicians (**4**) in NASA's Launch Control Room at the Western Test Range. This facility is located on the Van-

THE TRACKING NETWORK

How NASA keeps tabs on satellites and spacecraft

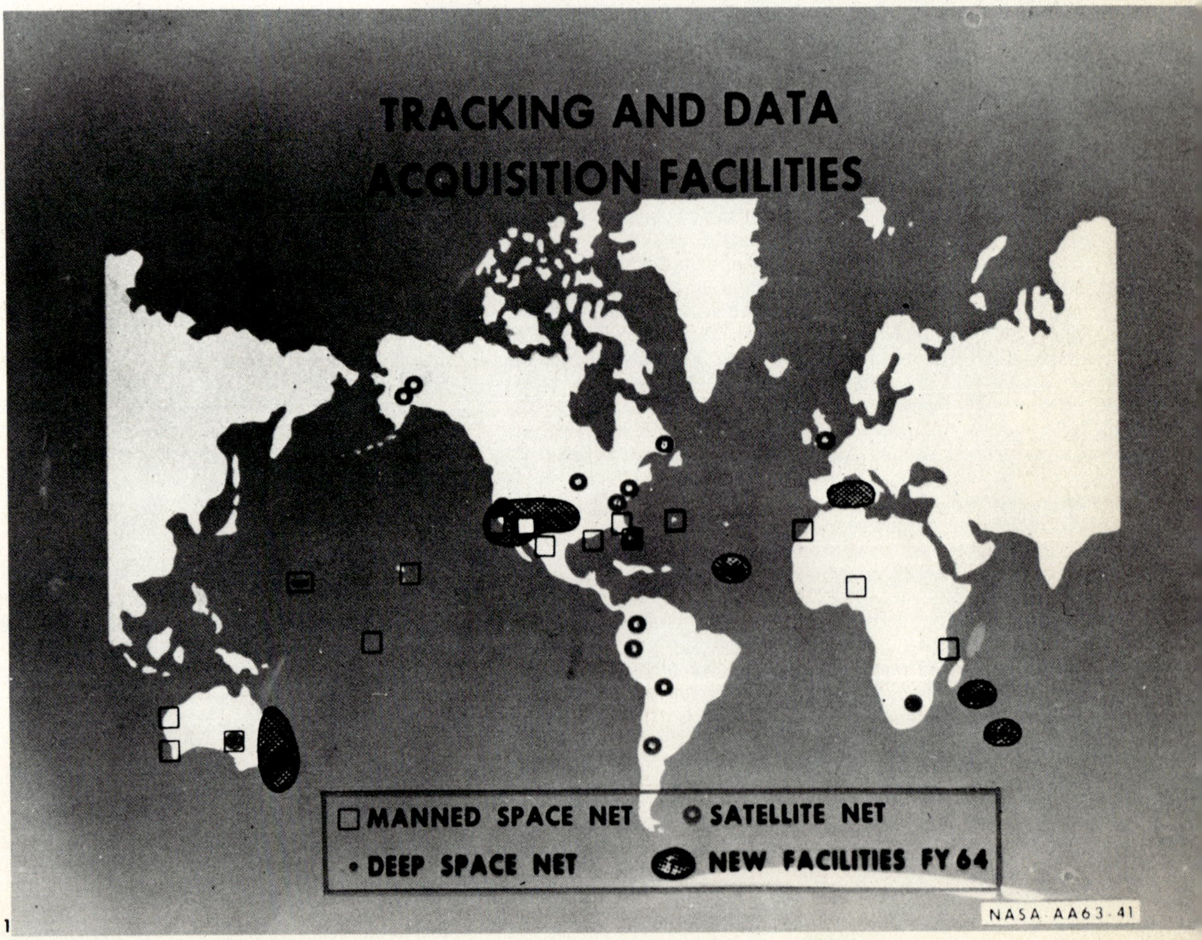

denberg Air Force Base at Lompoc, California. Vandenberg's launchings, unlike the widely publicized operations at Cape Kennedy, usually do not specify a vehicle's payload, as they are often military satellites put into polar orbit. Nevertheless, the operations at Vandenberg are part of the vital tracking network, as are those at the Pacific Missile Range, Point Mugu, California. A technician at Goldstone keeps an eye on one of the banks of high-speed digital recorders (**5**) as data comes in from the deep space probe Mariner 9. A new digital wideband communications line from Goldstone to the Space Flight Operations Facility at the Jet Propulsion Laboratory, Pasadena, California has a capability of 50,000 bits per second and allows real-time transmission of most data from a spacecraft. Another capability of Goldstone is daily television reception and transmission of pictures obtained and stored by on-board cameras in the spacecraft. This function is vital, since transmission is intermittent when a spacecraft is in orbit around a distant body. Goldstone's transmitter room (**6**) is the link that connects the 64-meter antenna to the Jet Propulsion Laboratory in Pasadena. During a manned space flight, Goldstone is the primary receiving station in the western United States, and the biggest in NASA's network. Precise positional information on a space vehicle is relayed simultaneously to Mission Control in Houston, Texas, to JPL in Pasadena, and other necessary links in the network. Decoding telemetered data is performed at the JPL facility.

THE TRACKING NETWORK

Tracking a space vehicle, whether manned or not, is an important function. This is true because knowing exactly where a space craft is at any particular moment allows its location to be fixed in relation to the earth, the sun or the moon. This knowledge permits the Mission Control technicians to receive and send accurate speed and guidance information, to ensure safety of the crew if the craft is manned, or even possibly to alter the spacecraft's flight trajectory if the vehicle should happen to be off course.

Among the smaller tracking antennas used in the global NASA network is a 40-foot parabolic antenna (**1**) at Mojave, California. A parabolic reflector is shown being assembled (**2**) at the Philco-Ford plant in Palo Alto, California. An operations room is seen at the tracking station in Fucine, Italy (**3**) where two operators are seated at a console which controls the antennas mounted outside. Note the chronograph clock at the left of the operator. This is an extremely accurate instrument of great value to trackers

wherever time precision is of utmost importance. Because spacecraft travel at such enormous speeds, their locations change considerably in fractions of a second. Next is an aerial view (4) showing the two 85-foot diameter antennas of the NASA station near Rosman, North Carolina. This station is one of the seventeen which form the network known as STADAN (Space Tracking and Data Acquisition Network). These stations locate and receive telemetry data from unmanned satellites, but are not involved in two-way communication with manned spacecraft. Tracking bases are often located in remote and unlikely places, which gives them freedom from the electronic interference caused by cities. The Mojave Desert in California is quite unlike the scene (5) which depicts technicians doing some repair and maintenance work on the tracking installation at Kano, Nigeria, which is in West Central Africa. Another tracking station is located at Malindi, Kenya, which is in East Africa along the Indian Ocean. Some tracking equipment is operated manually, at least partially so. Here (6) we see a technician seated at an optical control beside the dish-shaped antenna. Tracking is the continuing act of keeping in touch with flying or orbiting objects out in space. To do this it is necessary to have stations located around the world, stations with large radar antennas and electronic equipment. Such NASA stations are scattered in various locations around the globe, most of them fairly close to the equatorial path followed by satellites.

THE TRACKING NETWORK

Watching with one's eyes a moving person or object is as common as daylight. Watching a distant object through binoculars is quite similar. But tracking by radar is a different experience. There is yet another method, photography, which uses either static or moving pictures. The combination of camera, microscope, and telescope has developed into a science called *optics*. Researchers, engineers, and designers in the field of optics have conceived and developed marvelous long-range cameras to track missiles from blast-off to stage separation. Aided by these powerful cameras, we can watch from a safe distance the awesome controlled explosion at blast-off, record it on film, continue filming the ascent of the missile, and even catch the actual separation of the first stage booster and its dropping away from the second stage. With a Saturn V rocket, this first-stage separation occurs at a height of 33 miles. This powerful, sophisticated camera (**1**) is located at Melbourne, Florida, which is about 25 miles

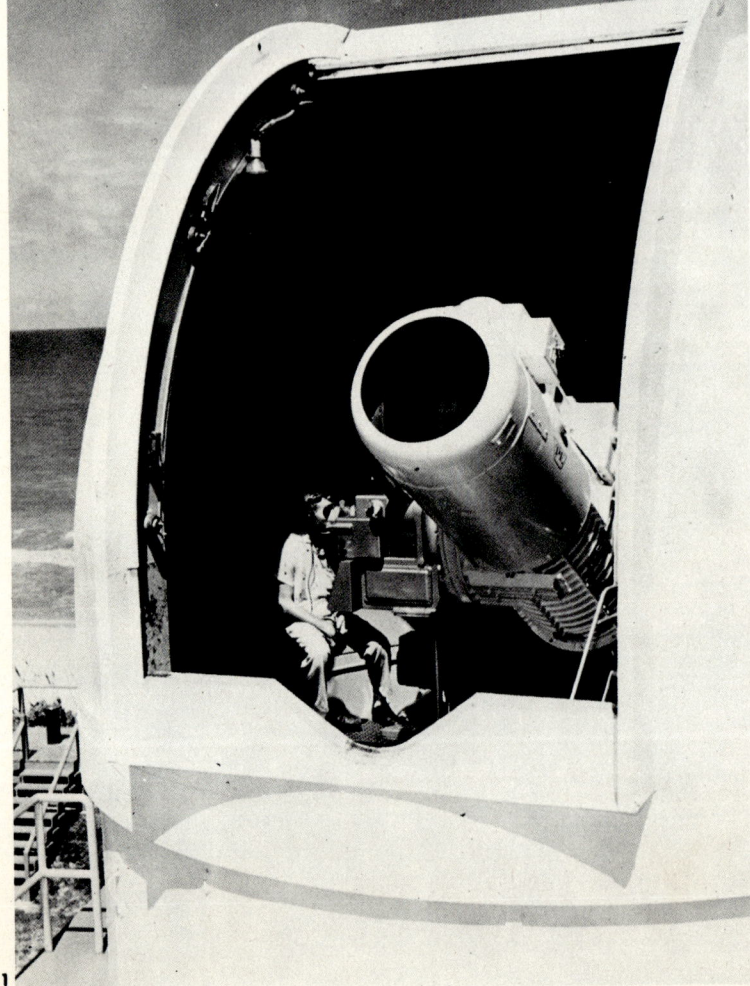

south of Cape Kennedy, and is capable of following and filming a missile as far up into the sky as 40 miles. The operator is peering into a viewer so he can guide the elevation of the camera, thus keeping it on target. Complicated test equipment is essential to flight in space. Stationed at Bermuda in the Atlantic is one of the hundreds of test engineers (**2**) plugging his tell-tale probes into a big informational switchboard. When distances between land bodies are a thousand miles or more, it becomes necessary to place ships at convenient locations on the oceans, and thus shorten the gaps between tracking stations. The large tracking ship (**3**) seen here is docked for service, but it is on station at sea when serving as a tracking station. It is equipped with a multitude of masts that bear aloft the needed electronic and radar gear. Some tracking takes place in cold terrain (**4**). The antenna is an upward spiral, somewhat like a circular stairway, with electronics-packed support trailers in the background. Thanks to systems developed for space missions, tracking by satellite is now commonplace. The weather satellite (**5**) relays to the station the presence of icebergs, which then can warn ships at sea to prevent another disaster like the sinking of the *Titanic*. Living objects can also be tracked as they roam their natural habitats. An elk (**6**) is shown wearing a transmitter put on by wildlife officials, which sends continuous signals to a satellite. This information is sent to a tracking station, and thus do scientists learn about animal migrations.

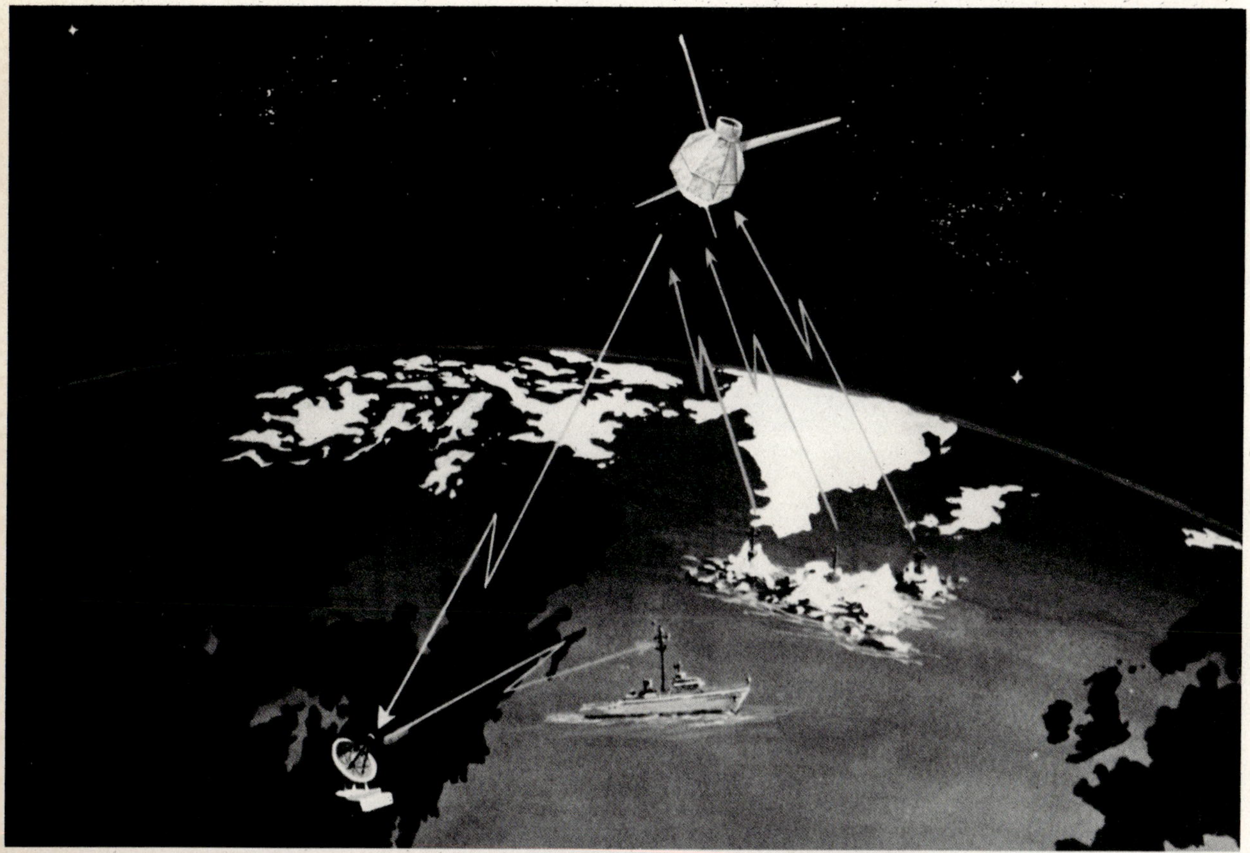

MAN IN SPACE VOL. 4/11

THE COMMAND & SERVICE MODULE

A guided tour of the Apollo crewmen's home in space

Originally, NASA studied two possible methods for sending men to the moon and back. The first utilized a very large launch rocket that carried the landing vehicle directly at its upper end. The crew would remain in the landing vehicle during the entire mission from launch to lunar excursion to return. While such a mission profile had much to commend it for a straightforward approach, the development of a suitable launch vehicle would have presented many difficulties from the aspects of size and cost. This study was abandoned in favor of the present Apollo configuration which utilizes separate space vehicles, or "modules," for separate tasks. There are three of them. The *Lunar Module* (LM), detailed in a separate chapter, is used only as a 2-man "space taxi" to reach the lunar surface from lunar orbit and return. The cone-shaped *Command Module* (CM) is the home and control center for the entire 3-man crew, while the cylindrical *Service Module* (SM) provides propulsion and support functions for the CM. These latter two

are usually joined together and often referred to as one unit, the Command/Service Module (CSM). The prime contractor for the CSM is the North American Rockwell Corporation. A historic meeting occurs at the Cape (**1**) when the Command and Service Modules for Apollo 11 and 12 cross paths in the Manned Spacecraft Operations Building. The Apollo 11 CSM is on the gantry crane being taken out after its tests, while in the foreground, the Apollo 12 CM and SM have yet to be joined and checked out. This simplified exploded view (**2**) of the Service Module shows the arrangements of its framing, tankage and propulsion. The SM is assembled at North American's Downey, California plant. Here (**3**) is the hull of SM #007. Vertical wall supports divide the area between the fore and aft bulkheads. In the central cavity, where the technician is standing, will be placed helium flasks. During assembly, the SM undergoes constant checks under superclean conditions (**4**). Electronics men (**5**) assemble the wiring harness on a special fixture, before tucking it into the CSM. The Apollo CM alone carries nearly 15 miles of electrical wiring. Here are two views (**6,7**) of the three electricity-generating fuel cells as installed in one of the six sectors of the SM. They supply most of the electrical power for the spacecraft and some of the astronauts' drinking water. With potassium hydroxide as an electrolyte and glycol as a coolant, each cell uses hydrogen and oxygen gas as "fuels," combining them to produce electricity (27 to 31 volts) and water.

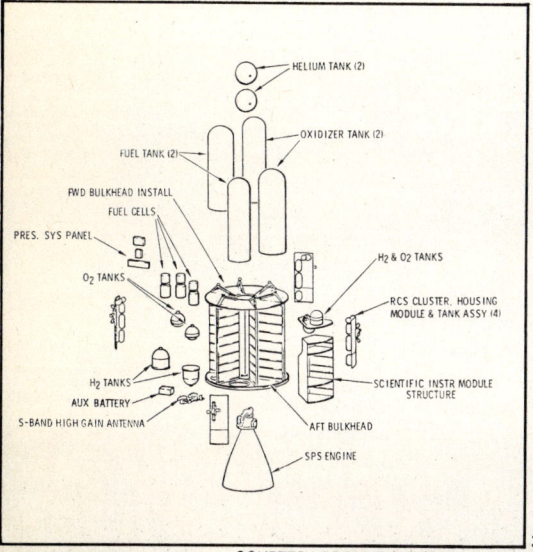
COURTESY ROCKWELL INTERNATIONAL

COURTESY ROCKWELL INTERNATIONAL

COURTESY ROCKWELL INTERNATIONAL

COURTESY ROCKWELL INTERNATIONAL

THE COMMAND & SERVICE MODULE

Of the 363-foot-high Apollo/Saturn launch vehicle, the only component that actually is recovered at the end of a mission is the 10-foot, 7-inch-tall Command Module which weighs 11,700 pounds at splashdown. Within it, the three-man crew remains during the greater part of a lunar mission; it is their home in space, providing their personal needs as well as most of their functions as astronauts. These line drawings (**1**) show the more important external features of a CM. A cutaway (**2**) reveals some of the more important internal components. The CM is an engineering and design miracle—it squeezes more complex equipment into a smaller volume than any comparable space vehicle, and leaves 78 cubic feet of living space for each of the crew. The CM is coated with an "ablative" heat shield, which burns away during re-entry into the earth's atmosphere, and in doing so, shields the crew from extreme heat. The central, or crew area of the CM (**3**), begins life as a precision-welded hull, joined on a frame or

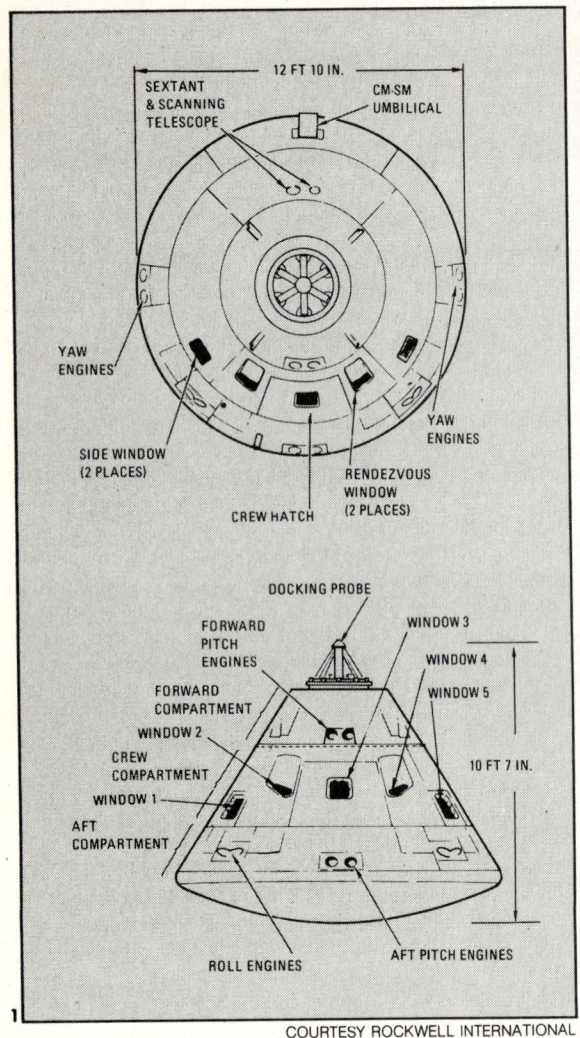

COURTESY ROCKWELL INTERNATIONAL

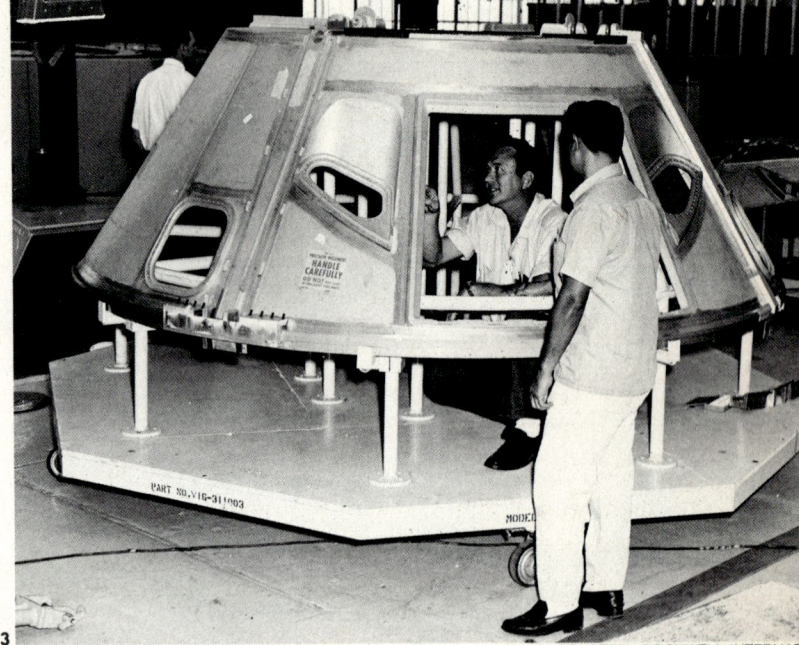

COURTESY ROCKWELL INTERNAT

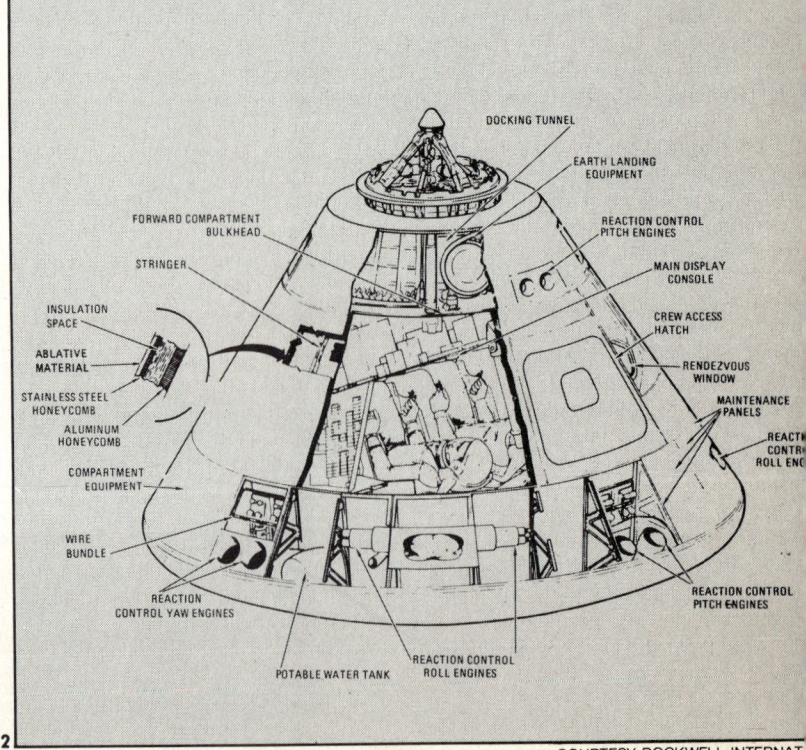

COURTESY ROCKWELL INTERNATIO

"jig". This view (**4**) shows a nearly complete CM on a large fixture at Downey where it is spun or tumbled, shaking out any foreign material that may be left from assembly. Note that the crew hatch, service panels and docking tunnel are open. During production, the CM is tested and checked repeatedly (**5**). In fact, the testing equipment is sufficiently sophisticated that a complete CSM has "flown" at least one "mission" before it leaves the ground; deep space does not offer second chances. During assembly (**6**), the outer hull of the CM is lowered into position over the inner structure. Note the wiring and valve groups arranged around the aft section. All assembly is carried out under very clean, rigorously controlled conditions. Considerable development has been undertaken on the CM, using principles originated on the Mercury and Gemini capsules. One prime consideration is the CM's ability to withstand extreme heat and extreme cold—at the same time. In space, the side facing the sun becomes superheated, the opposite side growing extremely cold. This heating cell (**7**) duplicates such conditions. It is shown in the open position with the heaters on; the opposite half of the cell is literally a huge freezer. (**8**) Splash! This test tower with pool allows engineers to test the effects of splashdown impact. The 143-foot tower drops a fully instrumented CM at controlled speeds and records data with tapes and oscillographs. The design impact speed of the CM is 19 miles per hour, and it can right itself with special balloons if it capsizes.

COURTESY ROCKWELL INTERNATIONAL

COURTESY ROCKWELL INTERNATIONAL

THE COMMAND & SERVICE MODULE

The crew's command and guidance center must also serve as their living and storage quarters throughout an entire mission. With the crew couches and main display/control panel removed, this line drawing (**1**) shows the interior compartmentation arrangement of the Command Module. All necessary equipment, particularly the environmental control components, wiring, batteries, and the navigation station are modularized in groups around the CM's bulkheads. A photographic view (**2**), looking towards the Left Hand Intermediate Equipment Bay, shows one of the docking windows at center. Above it, reaching across the upper part of the photo, is the main control panel. The cabin couches are removed, but their shock-absorbing hangers are seen hanging in position. To the immediate left of the window is the Crew Optical Alignment Sight, while at the base of the window is the T-shaped cabin pressure relief valve handle. In flight, the crew's feet face to the right of the picture. This diagram (**3**) indicates the loca-

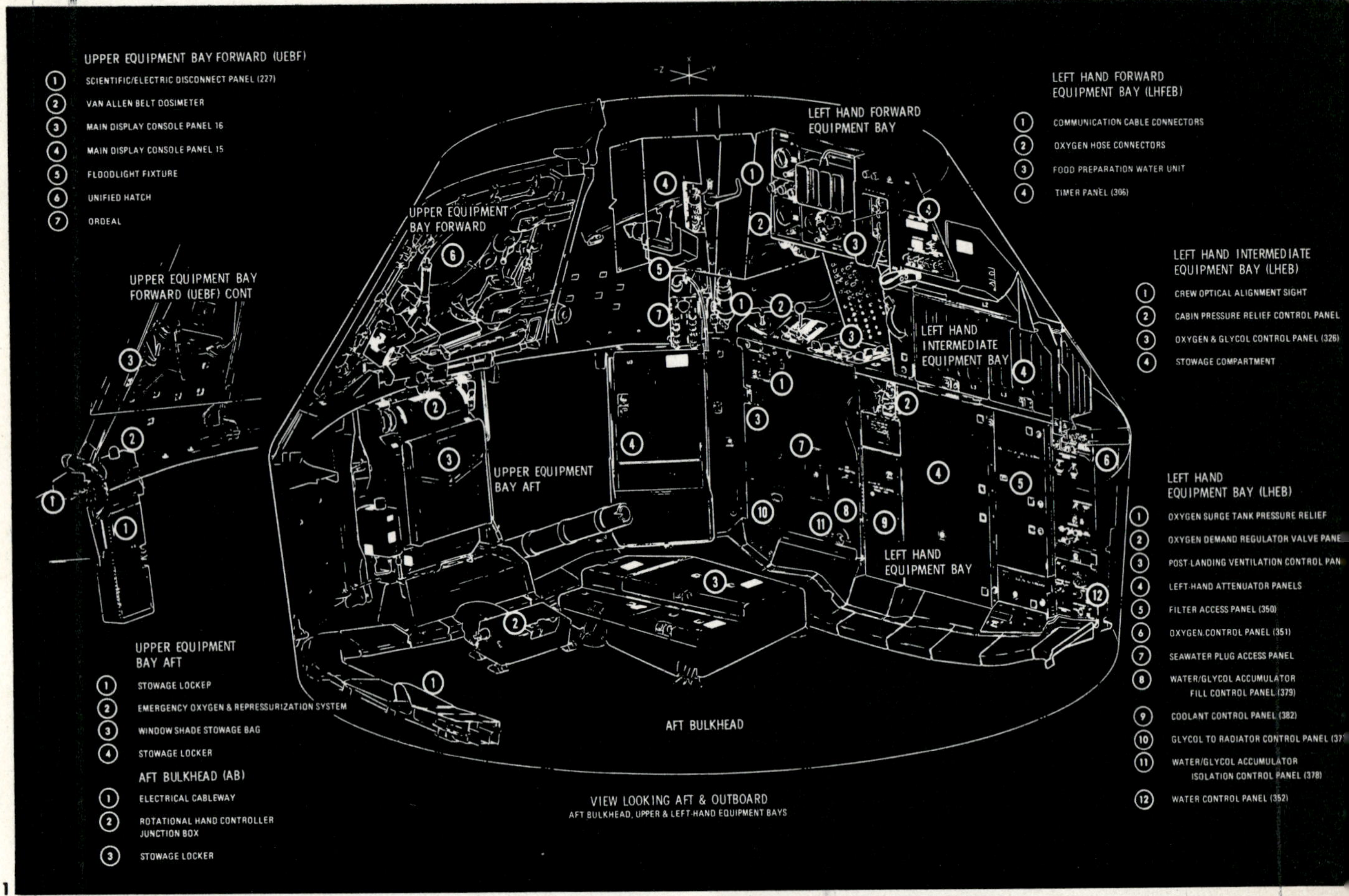

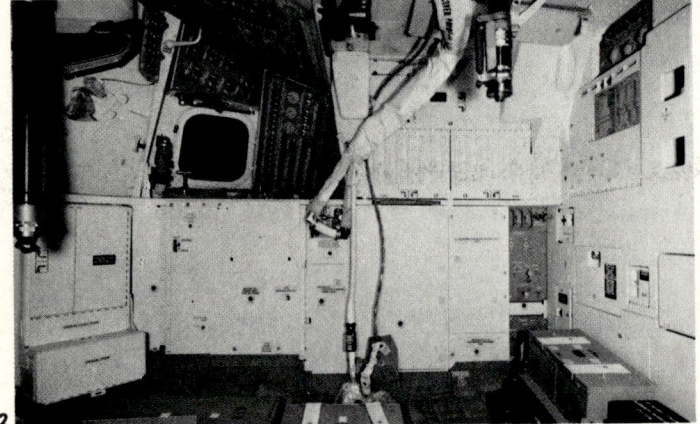

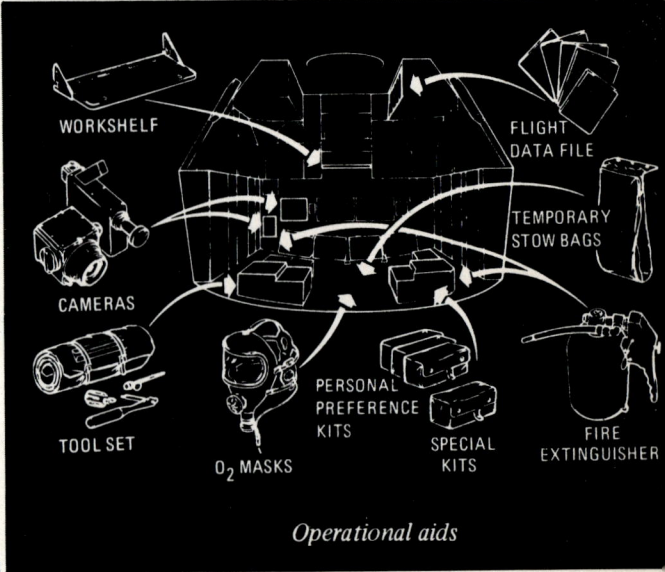

Operational aids

tion of the various flight and emergency operational aids. An extensive flight data file is stowed for both the LM and CM pilots. They contain the overall flight plan, mission log, landmark maps, star charts, and subsystem data. A tool roll containing a set of special hand tools is stowed on the aft bulkhead; placards on various panels indicate which tool is to be used where, and in which direction the fastening device is to be turned or manipulated. A 70mm Hasselblad and 16mm Maurer movie camera are part of the CM complement. The oxygen masks provide the crew with a quick source of breathable air should the interior environment become dangerous (smoke, gasses, etc.) while they are unsuited. With the couches and restraints in position, the CM cabin (4) becomes a bit more crowded! This view is taken looking through the access hatch and the seats are in their extended position. In the foreground are the rear suspension tubes for the couches. Overhead is the main switch and control panel. On the left-hand seat suspension tube is one of the rear view mirrors that aid the crew in operating and observing certain instruments. The CM pilot can unlatch and open the side hatch in three seconds using this ratcheting handle (5). It operates through gears on the 15 latches around the edge of the door, and was adopted after the Apollo 1 fire as one of the safety measures. The earlier design required some 90 seconds to release and open, and was a factor in the deaths of Gus Grissom, Ed White, and Roger Chaffee.

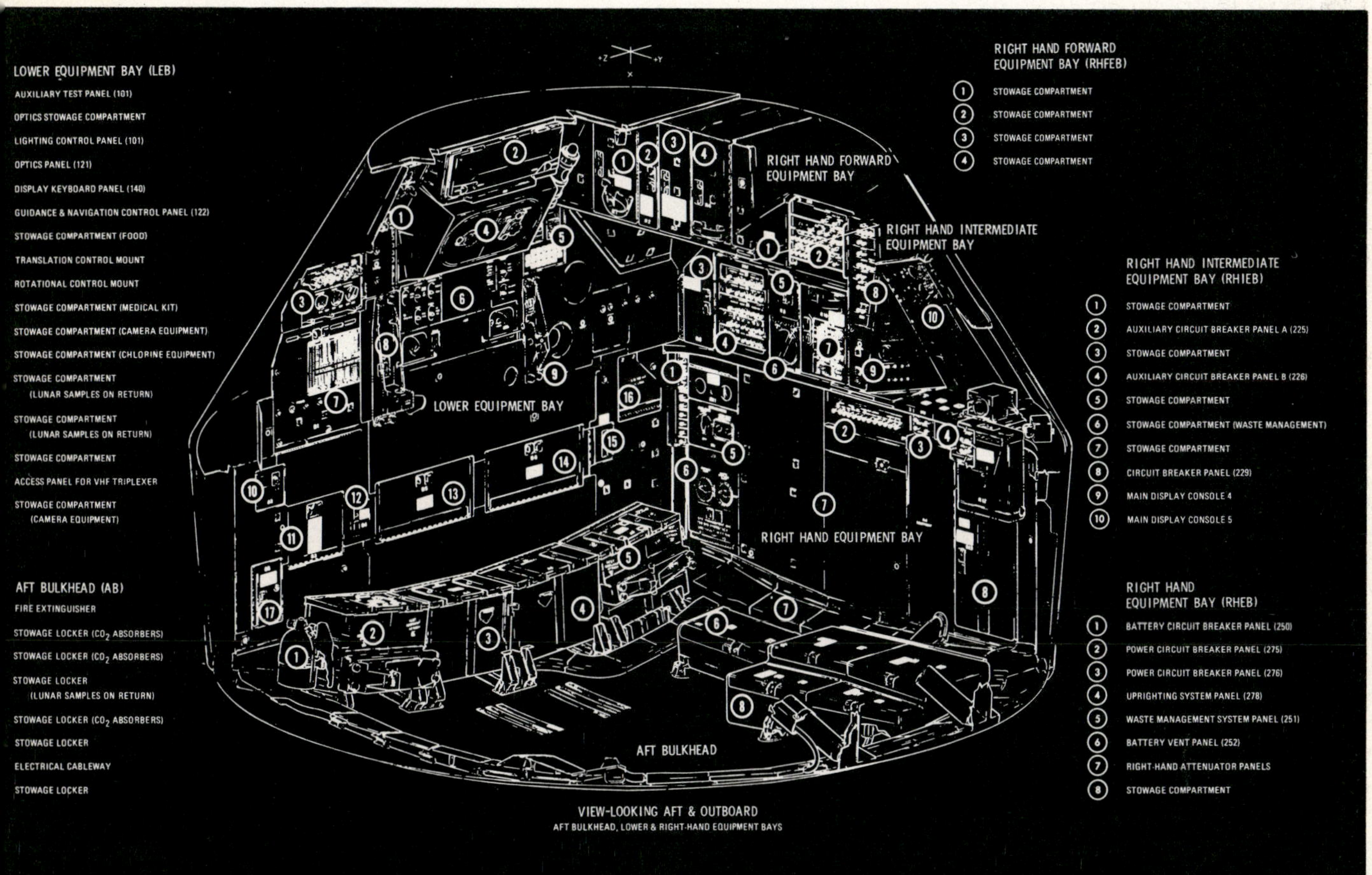

Arrangement of equipment in interior of command module

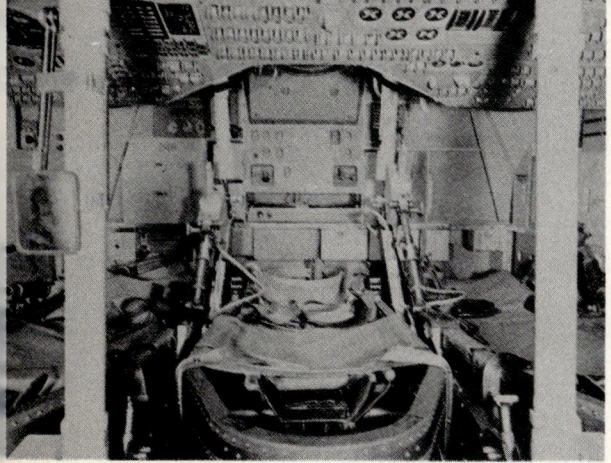

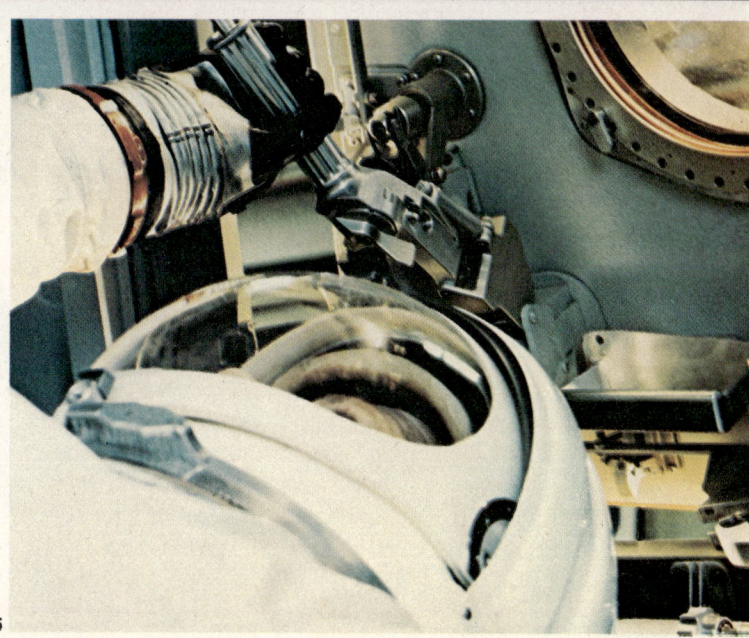

THE COMMAND & SERVICE MODULE

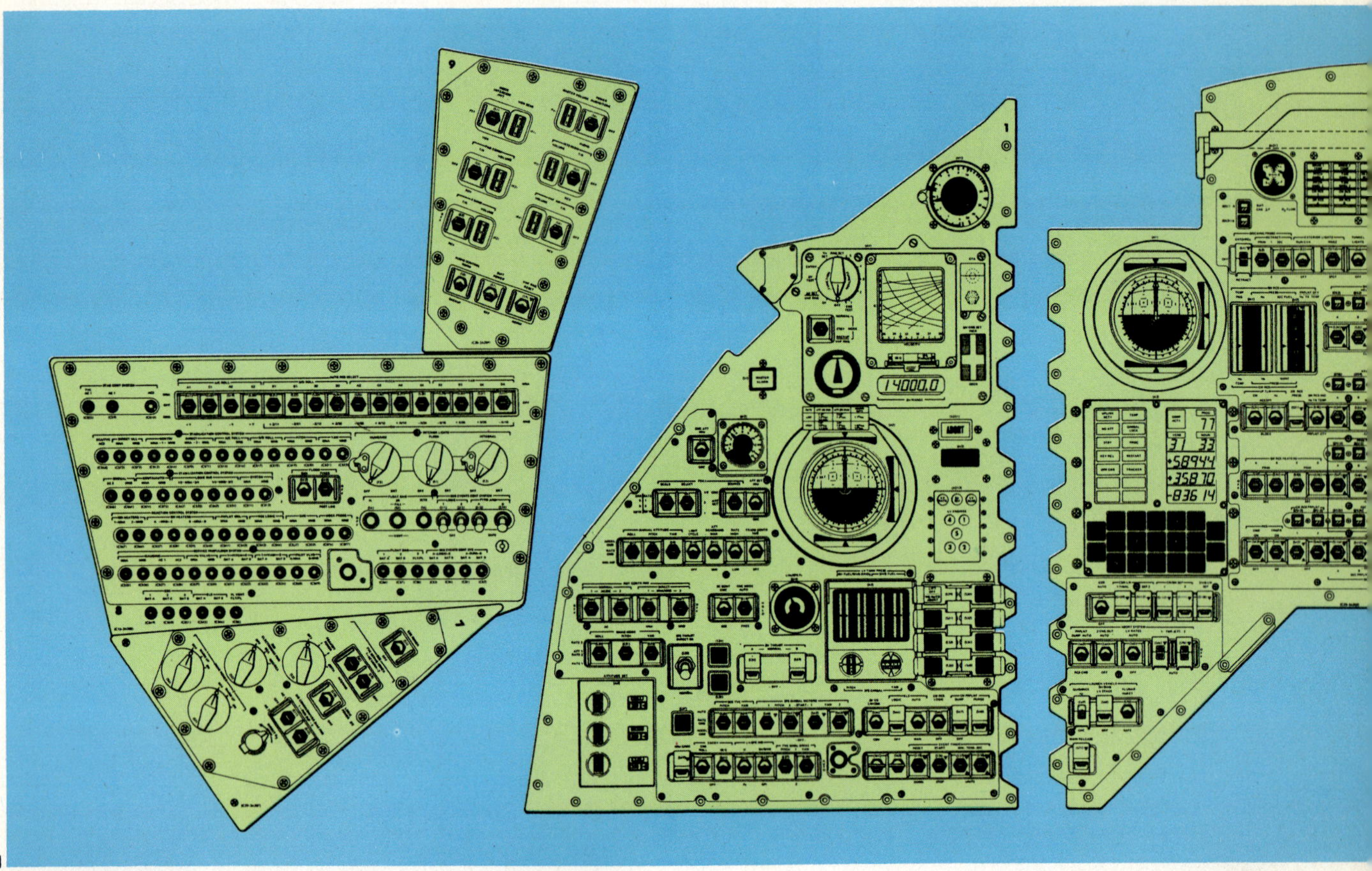

Lying across this spread is a drawing of the main display console (**1**). Nearly seven feet long and three feet high, the console is the crew's control center and monitoring area for the entire spacecraft. The diagram (**5**) shows the grouping of the various controls. All primary functions of the spacecraft are either controlled and/or monitored from this console. Numerous other controls are located elsewhere in the CM, notably the Guidance and Navigation Station, but they are considered to be secondary in importance. Each of the three crew members has a certain area for which he is primarily responsible on the console, but all are fully trained to monitor or operate any of the instruments. Every vital system and control is redundant; that is, they are equipped with warning monitors for malfunction or improper crew input. Redundancy also allows a secondary system to control the necessary function without interruption. During a fit and function test at North American's Downey plant (**2**), astronaut David Scott stands in the

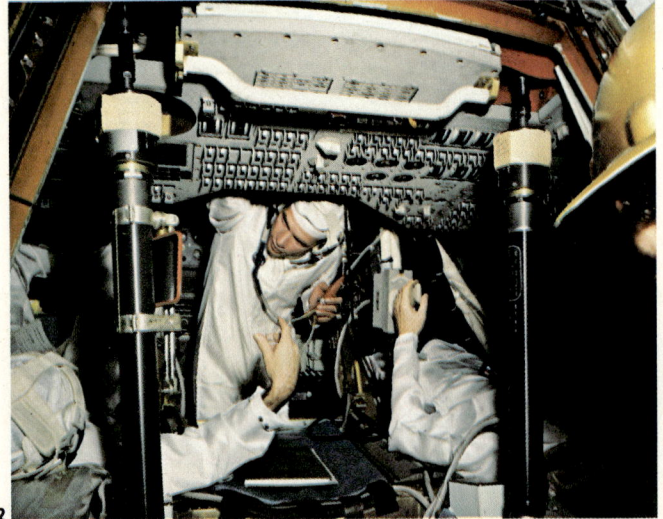

docking tunnel of Spacecraft #101's Command Module. Behind him is the Guidance and Navigation Station, while at center is the main console. Flanking Scott are astronauts Jim McDivitt at left in the commander's couch, and Russell Schweikart at right at the LM pilot's post. Schweikart's left hand is grasping the spacecraft's rotation control stick. This unit, of which there are two, is integrated with the gyros and propulsion system, permitting angular control of the spacecraft around its axes. It is also used for manual thrust control in pitch and yaw conditions. Scott is holding part of the communications line to his headset. The Apollo 8 prime crew (3), fully suited, train in the Apollo Mission Simulator at Kennedy Space Center. This photo looks towards the side access hatch of the CM from the standing area below the docking tunnel. Astronauts are fully suited during critical flight phases such as launch and re-entry, but may remove the suits during flight, since the cabin is pressurized to 5 psi. Their suits are connected to the CM's oxygen system with the hoses at center. Directly above them is the main console. This is the docking tunnel (4) through which the commander and LM pilot have access to the Lunar Module. In this view, the center couch has been retracted, but the two side couches are in flight position. The crewman at right has his hand on the ECS Indicator switch, which monitors glycol coolant flow in the Environmental Control Subsystem. The docking tunnel is approximately 30 inches in diameter.

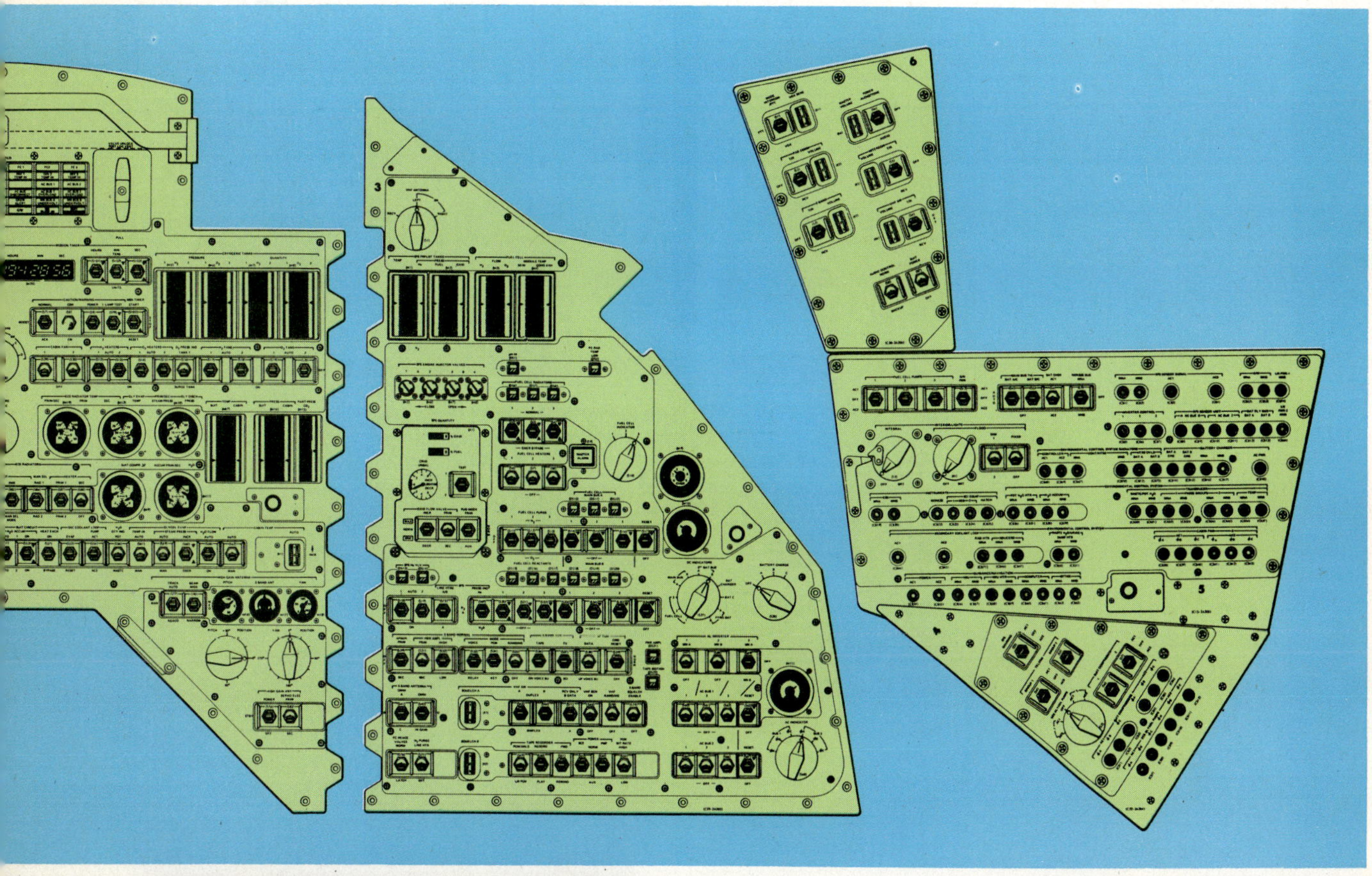

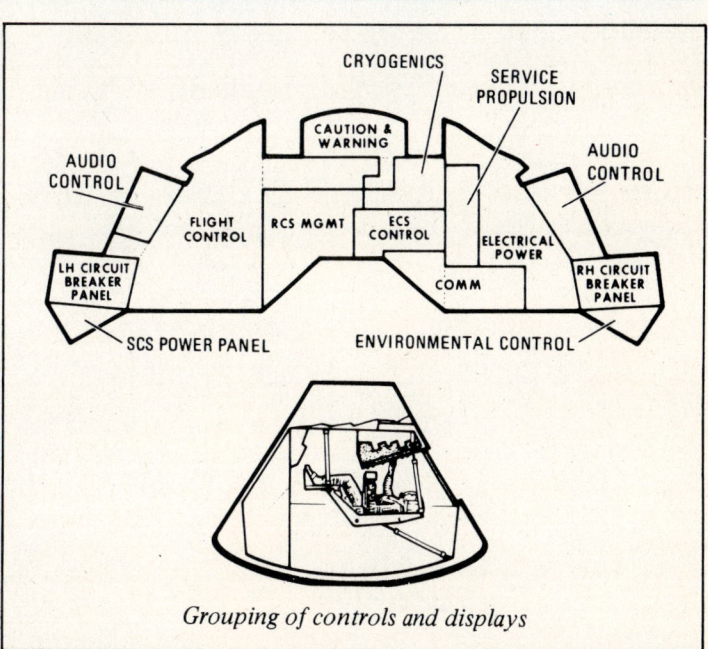

Grouping of controls and displays

THE COMMAND & SERVICE MODULE

One of the most critical components of the spacecraft is its Guidance and Navigation Subsystem. In the Command Module, the navigation equipment is located against the bulkhead in the stand-up area beneath the docking tunnel. In this position, it is accessible to the CM pilot, who normally operates it. A functional mockup (**1**) shows its position in the CM. The station has its own telescope for coarse alignment on ground or star landmarks, a sextant for actual "fixing" on the predetermined landmarks, and a computer (**2**). The computer is able to solve inflight or orbital course changes from inputs directed to it by the operator. Space navigation is more complex, because the Apollo spacecraft must navigate through three-dimensional space; it is not close to the earth at all times such as an airplane or ship. The Launch Escape Tower (**3,4**) is shown in section and in position on the CM. It can blast the CM free of the rocket vehicle at any point from launch pad to the time of 30 seconds after second-stage ignition.

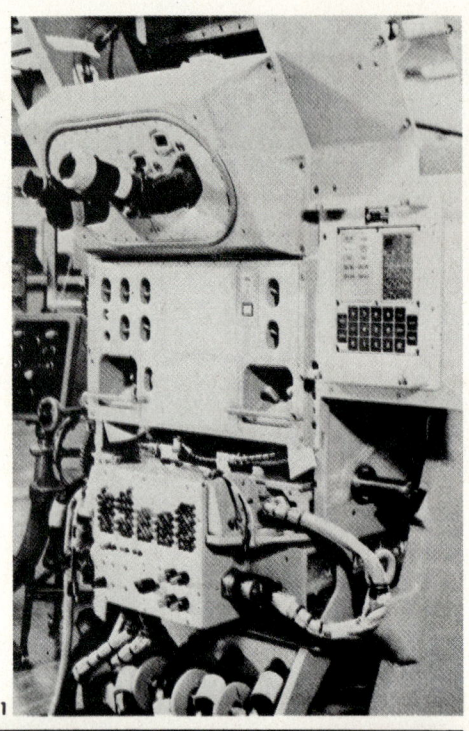

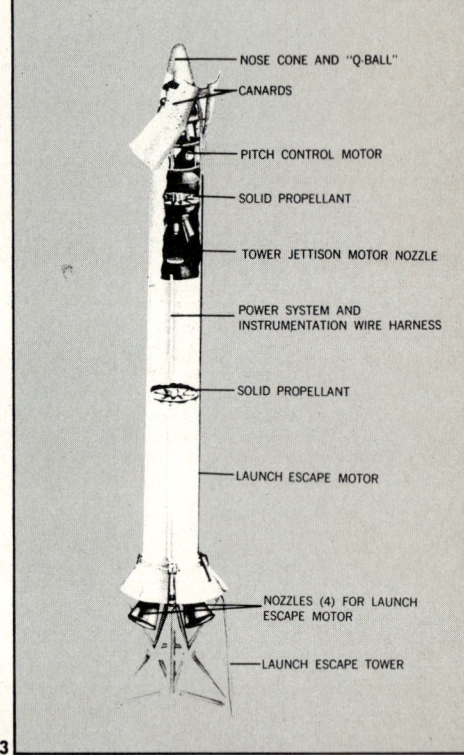

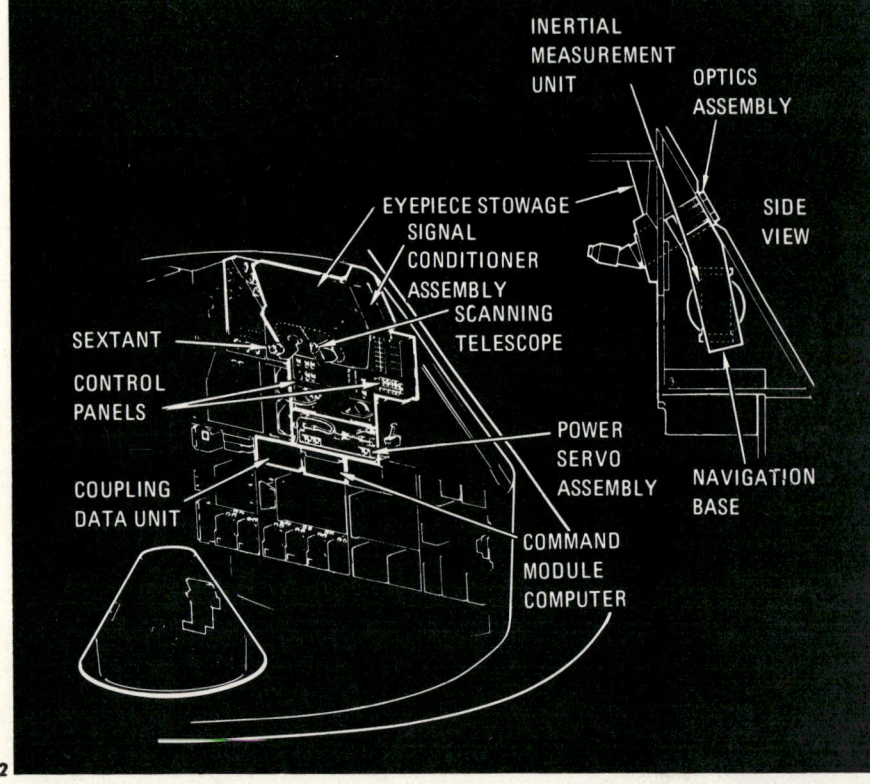

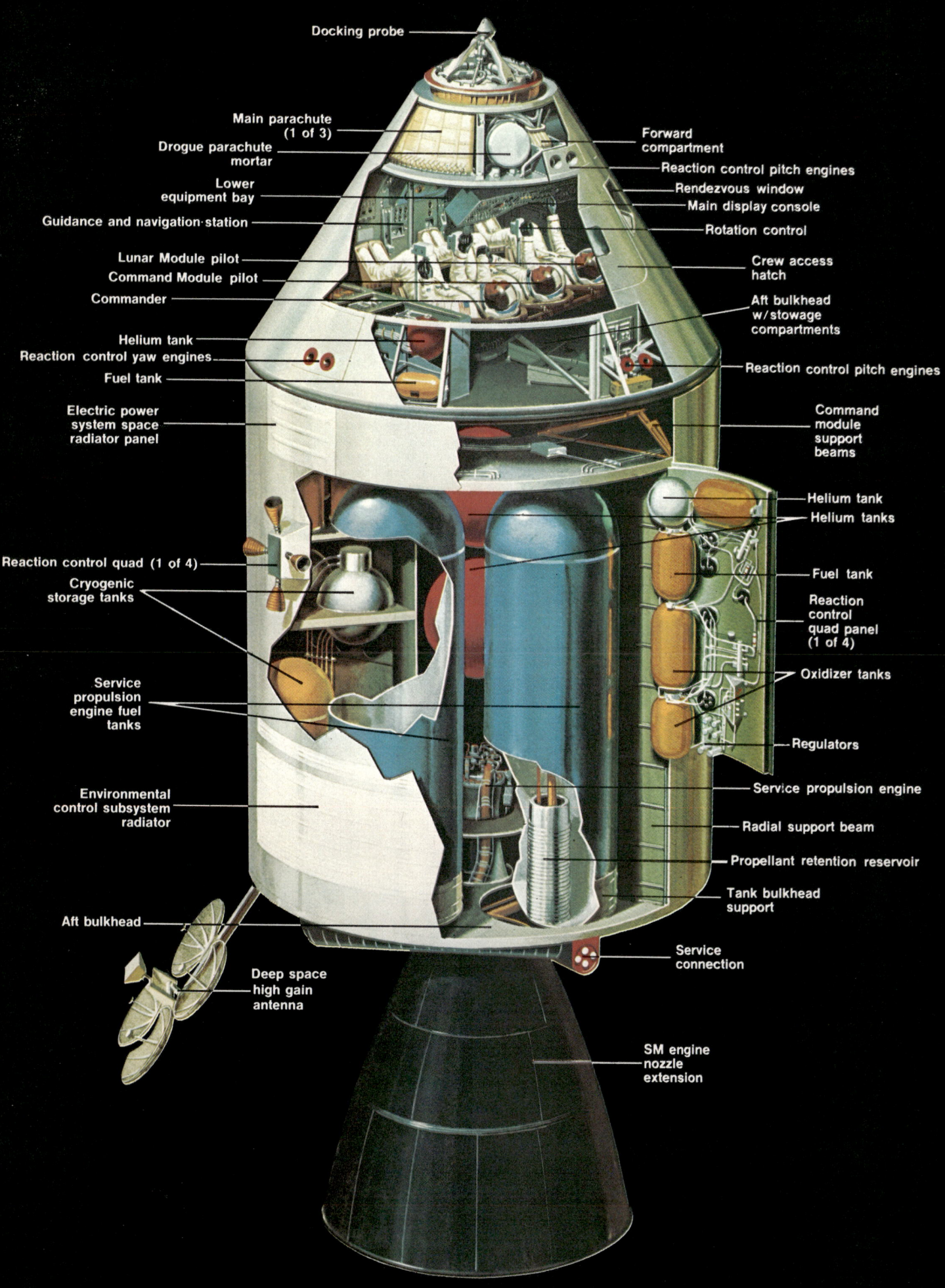

APOLLO 7

Rehearsals for a manned flight to the moon are about to begin with the scheduling of Apollo 7 for the fall of 1968. Although extensive testing of man's ability to survive and function in weightlessness has been carried out countless times, the real test of a lunar flight will come only when humans are launched into space. Until the first manned Apollo spacecraft is successfully launched and brought back to earth, no one can be absolutely certain that manned lunar flights will work, especially after the tragedy of Apollo 1. Apollo 7 is being readied for this task, the first manned space flight since Gemini 12, almost two years ago. Navy Captain Wally Schirra will command, making him the first astronaut to have flown Mercury, Gemini, and Apollo spacecraft. He will be teamed with Air Force Major Donn Eisele and civilian scientist-pilot Walter Cunningham. The astronauts have been assigned several complicated chores to perform in orbit, including a docking with the S-IVB second stage, simulating the docking with the Lunar Module future Apollo flights will carry. Watched with special interest will be the crew's evaluation of the safety modifications made to the Command Module as a result of Apollo 1. This is the kick-off in the race to the moon.

WALTER M. SCHIRRA, JR.

SPACECRAFT COMMANDER/ This flight makes Navy Captain Schirra, the crack pilot of *Sigma 7* and Gemini 6, the first astronaut to fly in all three types of American manned spacecraft. Since all other of the "original seven" are no longer active astronauts except Gordon Cooper, he will probably remain the only one with this distinction. Following the historic rendezvous with Gemini 7, he received an Honorary Doctorate in Astronautical Engineering from Lafayette College, and his interests in outdoor sports continue unabated.

DONN F. EISELE

COMMAND MODULE PILOT/ Born on June 23, 1930, in Columbus, Ohio, Major Eisele graduated from the Naval Academy but chose a career in the Air Force. He earned an MS in Astronautics from the Air Force Institute of Technology in 1960. Trained at the Aerospace Research Pilot School in California, he became a project engineer and test pilot at the Special Weapons Center at Kirtland Air Force Base, New Mexico. Selected by NASA in 1963, he has more than 4000 air hours, and enjoys free time with his three children.

R. WALTER CUNNINGHAM

LUNAR MODULE PILOT/ Born on March 16, 1932, in the heart of America at Creston, Iowa, Cunningham considers Santa Monica, California, as his home town. After joining the Navy in 1951, he served as a fighter pilot with the Marines, accumulating 3900 flying hours; he is now a Marine reserve Major. With a BS and MS in Physics from UCLA, he is close to obtaining a doctorate in the same field. NASA selected him in 1963 while a research scientist at the Rand Corporation. The father of two, he relaxes with handball.

APOLLO 7
Pre-launch

Getting ready for an Apollo flight involves considerably more than jumping into a space suit, climbing into a spacecraft, and taking off for points unknown. Years of study and work precede every mission. The astronauts must be thoroughly schooled in physical sciences, and most of them hold several degrees in studies pertaining to space and space technology. In addition to their considerable experience and expertise as pilots, the Apollo 7 crew holds degrees in science, physics, and astronautics. Seated left to right (**1**) Donn Eisele, Wally Schirra, and Walter Cunningham study an orbital map prior to their flight with Deke Slayton, Director of Flight Crew Operations. No, you're not being invaded by three mummies from outer space (**2**) in a second-rate science fiction thriller. This is an overhead shot of a Command Module simulator during a high-altitude test. On the left are the knees of CM pilot Eisele, while the "mummy" at right is Cunningham's left knee. (**3**) Eisele (foreground) and Cunningham are prepared

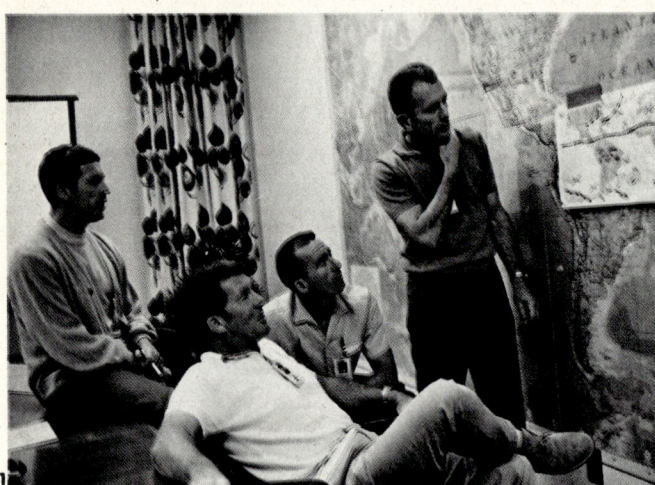

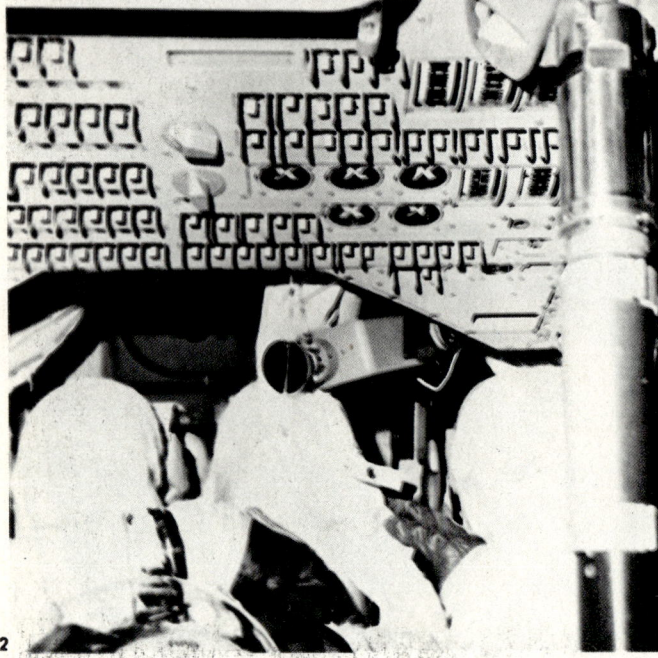

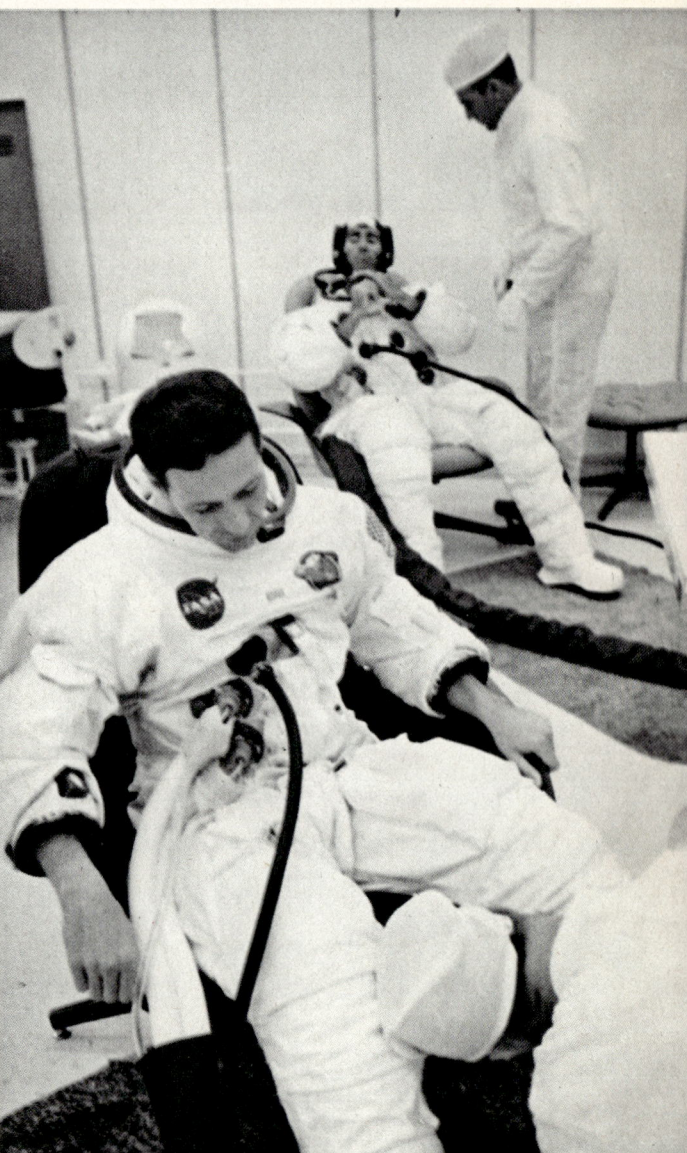

for another simulated flight in the altitude chamber at the Manned Spacecraft Operations Building at Cape Kennedy. Rub-a-dub-dub, three men in a tub (**4**); adrift in darkest Texas, the crew swaps sea stories during a quiet moment in water egress practice. They are afloat in the big water tank in Building 260 at Houston's Manned Spacecraft Center. (**5**) There is no fireplace out of camera range here, even though Wally Schirra appears to be warming his hands at some unseen source of heat; he is suiting up for a space vehicle emergency egress test at Cape Kennedy's Launch Complex 34. Looking very much like a monstrous metal baby in a giant incubator (**6**), Apollo training vehicle 2TV-1 stands in the Space Environment Simulation Laboratory in Houston. In it astronauts Joseph Kerwin, Vance Brand, and Joe Engle simulated a 7½-day flight, duplicating most of the Apollo 7 mission. Following all the training, launch day for the first manned Apollo flight finally arrives—October 11, 1968. The press gallery is filled with photographers for this historic occasion (following page). The countdown proceeds without delay, and ignition of the eight-engined first stage occurs right on schedule at 11:03 AM EDT, lifting the Saturn I B and its human passengers from Pad 39A into the blue Florida skies and then the blackness of space. The long-range eye of a telescopic camera peers deep into the heavens and catches the dramatic moment of first-stage separation at an altitude of 33 miles, followed quickly by second-stage ignition.

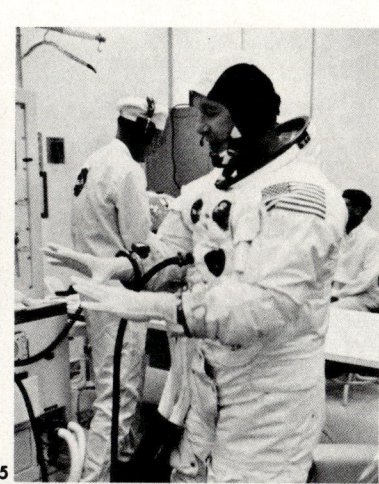

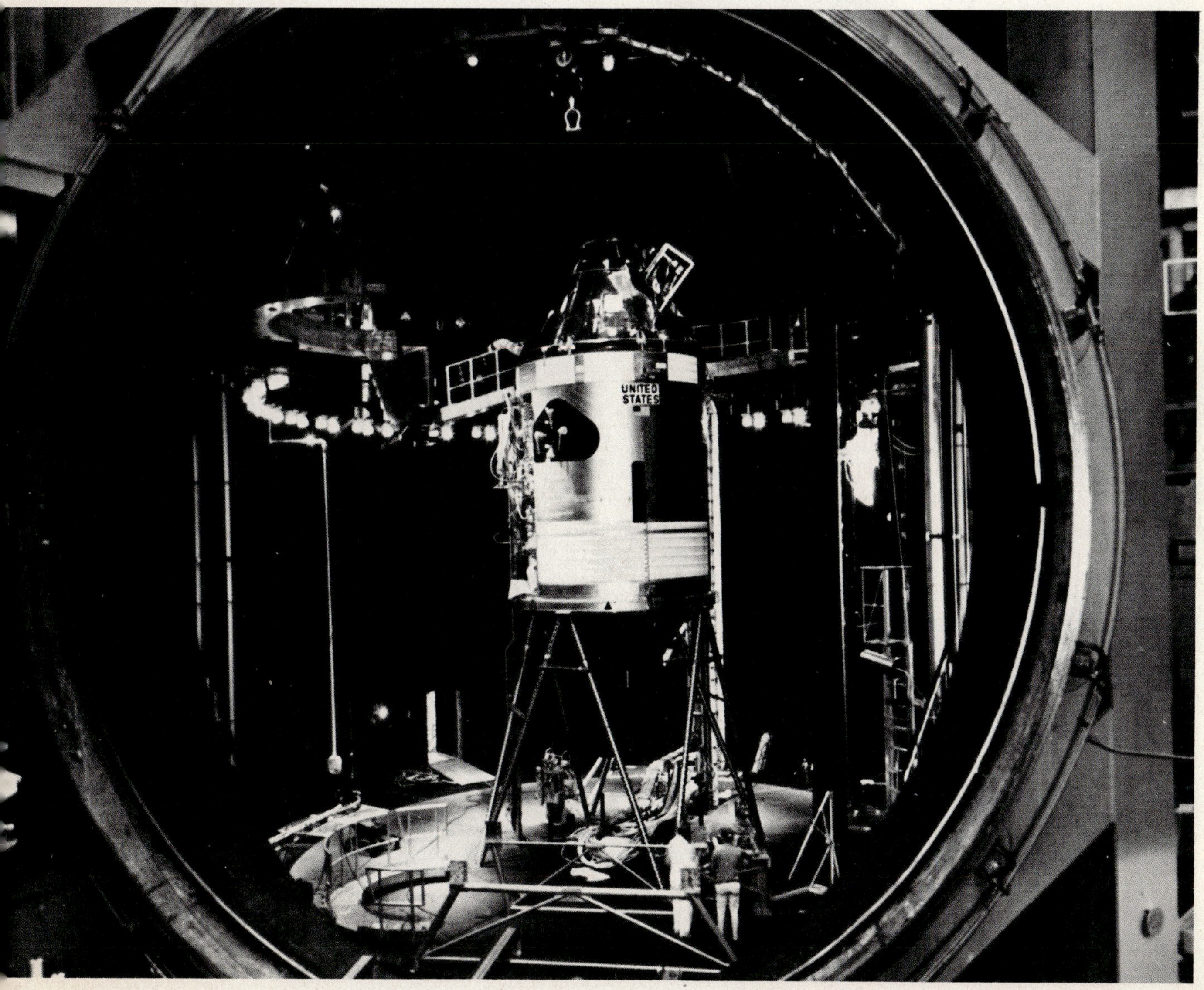

APOLLO 7
Launch

APOLLO 7 Mission

Several days into the mission, a stubble-faced Walt Cunningham (**1**) concentrates on the control panel inside the spacecraft cabin. A patch of blue sky, seen through a window, seems to perch on his right shoulder. Solar glare can be blinding in airless space, and here (**2**) Cunningham has donned sunglasses as he writes a report. Nine days in space, his stubble has grown into a respectable beard. Seeming to float right in front of his nose, but actually over by a window, can be seen a small cylindrical roll of film for one of the several 70mm Hasselblad lunar cameras on board. Astronauts on every mission seem to enjoy allowing small objects like this to wander around inside the cabin at times, probably because the experience cannot be duplicated on earth. These (**3**) are the Himalayas as they appear from an altitude of 130 miles, looking northwest along the Nepal-Tibet border. The world's dozen highest peaks, which reach a height greater than five miles above sea level, stand out in stark relief in this magnificent

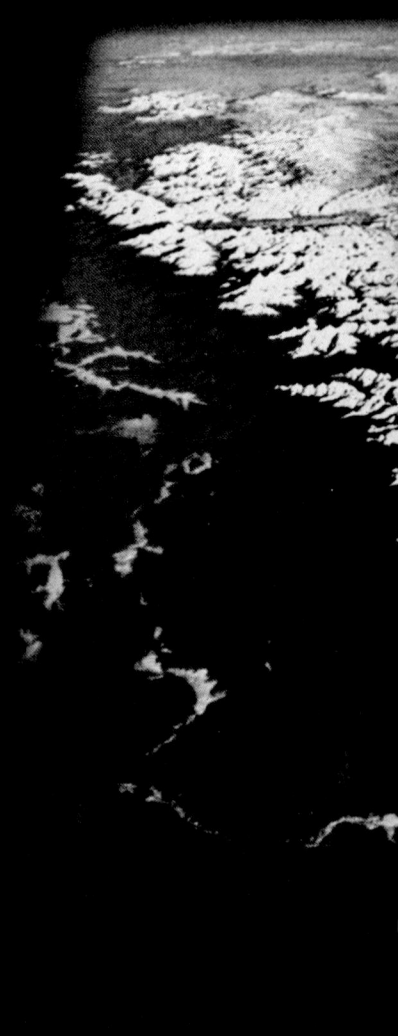

and remarkable view of "the roof of the world." Mount Everest, at 29,028 feet, is at lower center; 28,250-foot Mount Godwin-Austen (also known as K2), the world's second-highest mountain, can be seen on the central horizon, some 800 miles northwest of Everest; and in the lower right is Kachenjunga, the world's third-highest peak, which towers 28,208 feet and separates Sikkim from Nepal. The snow line is at 17,500 feet, and the lake-studded highlands of Tibet at the right are almost at that level; the river valleys of Nepal on the left are fed by these eternal snows. Donn Eisele (**4**) grins cheerfully for the camera during a pause in his duties aboard Apollo 7. That heavy growth of beard is the result of nine days sans shaving. No one has yet been able to figure out a way to prevent hair particles from floating away in a zero-gravity environment, possibly jamming some delicate instrument or electronic device, so astronauts do not yet shave during space flights. Judging from Eisele's smile, he doesn't seem to mind very much. (**5**) This may look like the stars of deep space, but it's really a close-up of debris on the outside of the spacecraft window. Almost all of this came from the exhausts of the various rocket engines, particularly the little attitude control thrusters mounted on the Command Module. (**6**) On its ninth orbit of the earth early in the mission, Apollo 7 passed directly over Kagoshima, Japan, yielding this startingly-clear view of the city and its surrounding bay area through the snowlike cloud cover.

APOLLO 7 Mission

Heading southeastward into the sun during its ninth day in orbit (**1**), Apollo 7 approaches the flat silhouette of Florida from across the Gulf of Mexico, and will shortly pass over its launch site into the Atlantic. Glinting in the morning sun can be discerned the swampy expanse of the Everglades, a remnant from a bygone age in earth's geological history, home to millions of primeval creatures oblivious to the passing of a spaceship above them. The onboard TV camera (**2**) catches Eisele (center) and Schirra (right) as they perform routine tasks during their first telecast, transmitted in their 45th orbit on October 14th. Notice that Eisele has buttoned the Command Module flight controller to his coveralls. In a takeoff on a popular TV show, Schirra signs off their first broadcast with this hand-lettered sign (**3**). The crew exhibits such humor during their telecasts that they become known informally to delighted viewers and TV commentators as "The Wally, Walt, and Donn Show." A mighty maelstrom in the Gulf of

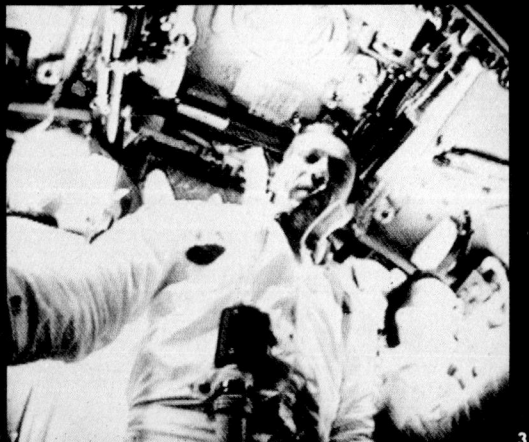

Mexico (**4**); 150 miles southwest of Tampa, hurricane Gladys swirls her cloudy formations counter-clockwise in an atmospheric ballet as incredibly beautiful as it is violent. The monster storm is over a hundred miles in diameter, but the crew of Apollo 7 is well beyond reach of its force or its sound as they orbit silently above, early in their seventh day. Perhaps the most disputed and fought-over land in human history, the Sinai peninsula (**5**), is seen between the Gulfs of Suez (left) and Aqaba as Apollo passes over the Red Sea. In the upper left is the Nile delta, while the Mediterranean is in the center distance. Mount Sinai, the traditional site where Moses received the Ten Commandments, is in the group of mountains at the lower end of the peninsula. High over the tortured terrain of Sonora, Mexico, the S-IVB stage (**6**) presents both a challenge and a dilemma. The job of the crew, particularly CM pilot Eisele, is to dock the tip of their CSM with the bright circular docking target, which simulates the same point on a Lunar Module. They approach closely, but fear that the slight impact of docking will swing those four big hinged doors inward (each one is 21 feet long), damaging the CSM. Schirra finally scrubs the docking attempt. The crew advises Mission Control that future LM compartment doors should be blown clear by explosive charges. Cunningham watches Eisele manipulate controls (**7**) during one of the mid-mission telecasts, while the seventh and last TV episode is concluded (**8**), "... and as the sun sinks slowly in the west..."

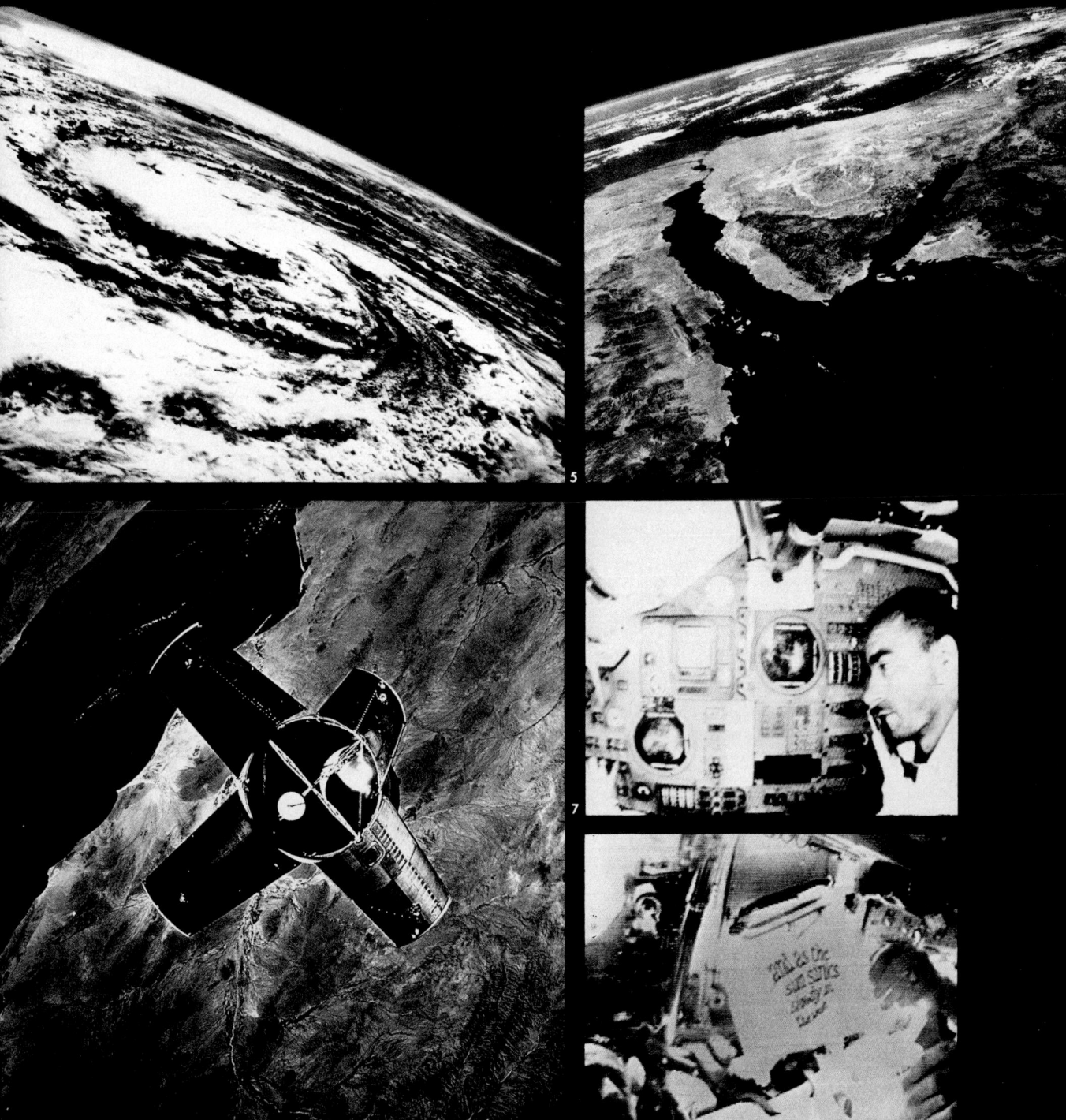

APOLLO 7
Splashdown

Although the seventh television transmission was their last, the astronauts still have plenty of work to keep them busy. The flight plan had been jammed with tests to be performed by the crew. Because the manned flight of Apollo 7 is innovative, Mission Control keeps suggesting that the crew make additional tests not previously included in the flight plan. Schirra becomes increasingly irritated by all these last-minute requests, and finally refuses to comply until each test can be thoroughly reviewed by the crew before it is acted upon, and Mission Control agrees to allow the astronauts to make their own decisions in this regard. There are no serious flaws in Apollo's performance. One of the three electricity-generating fuel cells starts to overheat, but this is cured by turning it off until it cools. Since the crew still has severe colds, Schirra wants them to come down without their having to wear space suits. Even though his concern about the crew not being able to clear their heads during the rapid pressure changes of des-

cent is valid, Mission Control insists upon the suits being worn, and a compromise is finally reached: suits must be on but helmets and gloves may remain off. The landing is smooth and uneventful; Apollo 7 splashes down ⅓ mile off target in the Atlantic Ocean at 7:11 AM EDT on October 22, 1968. Schirra's statement that "The mission went beautifully" is good news, and all concerned are delighted. Navy personnel aboard the prime recovery ship the USS *Essex* (**1**) greet the three space pilots with their own special welcome as their recovery helicopter nears the aircraft carrier. Pararescuemen (**2**) look into the open hatch of the Command Module after the astronauts splash down in the Atlantic Ocean. The three inflated bags atop the spacecraft enabled it to turn right side up after splashdown. Heavily bearded, the Apollo trio receives the traditional red carpet welcome (**3**) aboard the *Essex* upon completion of their 11-day earth orbital mission. The Apollo 7 spacecraft is recovered after splashdown and brought aboard the *Essex* (**4**) where technicians check it out. The spacecraft's scorched heat shield withstood re-entry temperatures of over 3000°F. The astronauts receive well-deserved awards at ceremonies in Washington, D.C. on November 2, 1968. President Johnson (**5**) beams with pride and pleasure as he accepts photos taken in orbit from Wally Schirra, as Eisele and Cunningham look on. Apollo 7 was a complete success, and paves the way for Apollo 8—a daring Christmas flight around the moon, the first voyage into deep space.

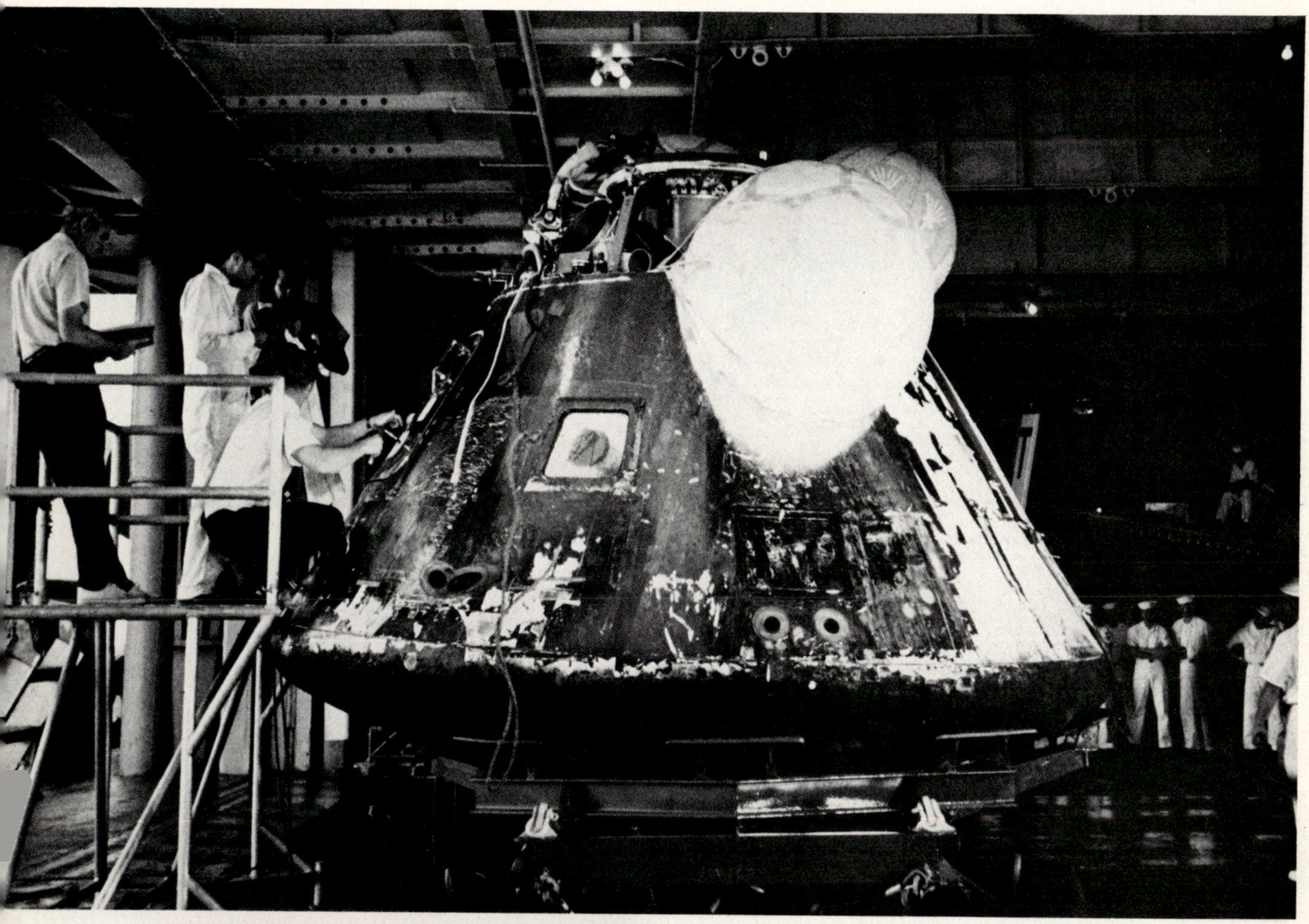

There was a time—before man learned how to hurl spaceships at the heavens—that any concept of eating in prolonged weightlessness would wallow in a sort of Buck Rogers fantasy. Proposals included aspirin-like nutrition tablets, artificial greenhouses, liquid diets, vitamin toothpaste, and a whole series of food substitutes practically guaranteed to de-humanize the most ardent space traveler. The concept of eating inside a spaceship, hurtling through a vacuum at speeds of thousands of miles per hour, was aggravated by the resulting zero-gravity. Eating at best would be a test of endurance; nothing like the pleasure man has always sought from this most necessary function. He must eat to survive. To that fact he will always remain in bondage. In a spaceship, there can be no crumbs. Water will not flow. Food size must be so small it is almost tasteless. All these food restrictions exist in missions which are among the most demanding mental exercises ever attempted. Historically, man has always looked forward to a good

THE RED, WHITE & BLUE

Breakfast, lunch & snacks ... the story of space food

meal after a hard day's work. Accordingly, an artist's conception (**1**) depicts an astronaut eating in flight—a non-required task in the first five Mercury shots—but of major importance in all space efforts starting with Gemini 4. In that flight, a 21-minute space walk by Edward White revealed that hungry-man appetites are not restricted to men on earth. NASA's efforts, therefore, now include checking a date-fruit cake (**2**) with calipers; and addition of special chewing gum (**3**) for the cleansing of astronauts' mouths after eating. A further evidence of NASA's determination that space food not be tasteless comes with examination of these strawberry cereal cubes (**4**) by nutritionist Patricia Cope and Robert B. Wheaton, manager of Life Support Services of the Whirlpool Corporation, which supplied much of the space food. Gemini meals and utensils (**5**) include scissors to open packaged foods, and a pistol-like probe to add water to freeze-dehydrated foods. A face cloth is added to each meal package. The probe (**6**) provides a correct portion of water so the mix can be properly kneaded prior to eating. Freeze-dehydration is a remarkable concept which enables foods to be stored almost indefinitely in a small size, yet be returned in taste and texture to original upon the restoration of water. Such Gemini foods are rigorously tested (**7**). NASA has strict specifications, and those foods accepted are sophisticated nutriments. We've come a long way in creating take-along meals since those early sailing ship diets of salt pork and hardtack.

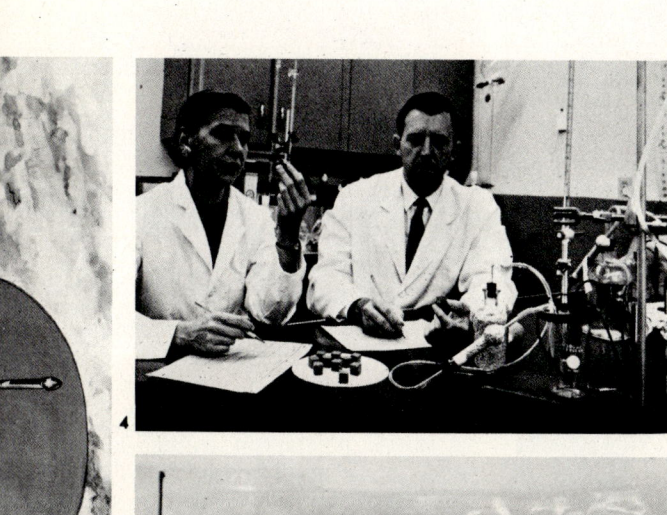
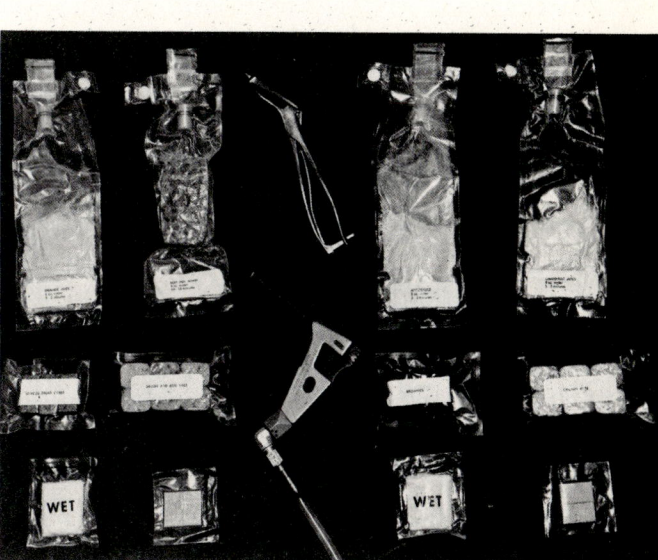
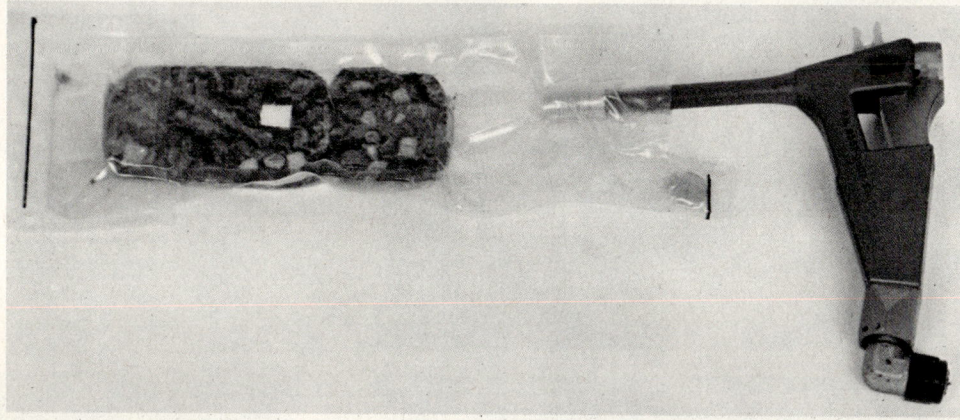
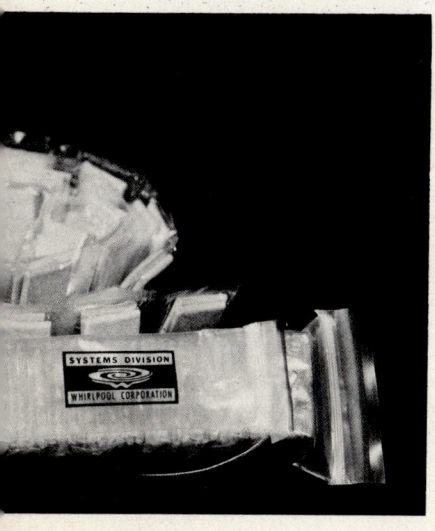
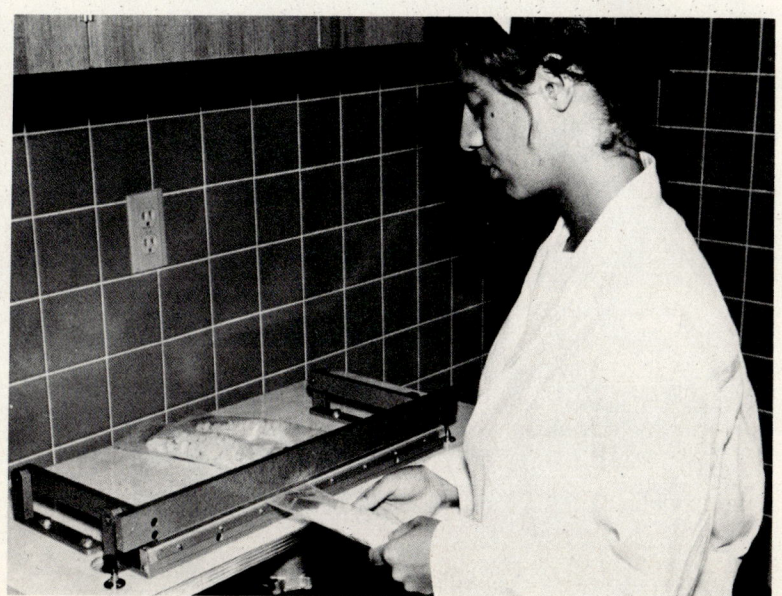

MAN IN SPACE VOL. 4/35

THE RED, WHITE & BLUE

Modern food preparation is a triumph of research and development. A great deal of time and effort is being spent in development of food for the astronauts. NASA is determined to further improve its specifications for a space diet of high protein, low waste items which retain the texture and flavor of the real thing. There is no doubt, of course, that the single most stubborn obstacle is the looming problem of liquids and solids in the weightlessness of space. NASA now knows that less energy is required to work in a weightless state, assuming the old bugaboo saying of "all things being equal," and a prescribed 2300 to 2500 calories daily seems reasonable. The astronauts have found it easier to eat in space than previously thought, but have advocated tasty food for some sort of psychological well-being. A diet composed entirely of semi-liquids might be nutritionally perfect, but without appeal. The problem is one of adding bulk, taste and texture to a meal, and of seeking ways to utilize a familiar utensil,

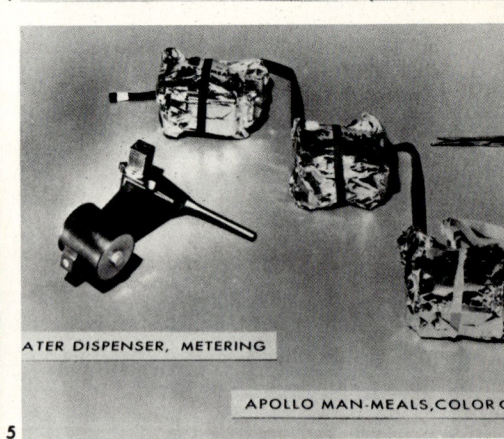

such as a knife or fork. The first space foods required acceptance of eating meals from toothpaste-type tubes. That, in turn, evolved to the present plastic pouch idea, perhaps the best-known spin-off from NASA food research. In its commercial variation this uses measured portions of vegetables in plastic bags, or becomes an "instant breakfast" by adding water or milk. The NASA plastic pouch is far more concentrated nutritionally, and far more expensive, as might be expected; there is more at stake. The ultimate goal in space food research is to approximate a home-cooked meal, a concept once thought impossible but now plausible. The conventional home meal looks quite different from its space equivalent (**1**); the latter uses the plastic pouch concept, complete with a handle (lower left) for water and metering attachments. The pouches (**2**) are not at all unappetizing, while the bite-sized sandwiches (**3**) are coated with a digestible gelatin to control crumbling. Developed for Apollo missions are these examples (**4**), featuring peas, orange drink and cocoa bags. Apollo "main dishes," called man-meals (**5**), are color coded and designed to be eaten in sequence. A treat awaits the crew of Apollo 11, with a package of beef and vegetables (**6**) that can be eaten with a spoon. Other Apollo meals include this breakfast kit (**7**), complete with orange juice and a toothbrush. The aluminum-wrapped space meals (**8**) contain complete daily menus, and are color coded: red for breakfast, white for lunch, and blue for snacks.

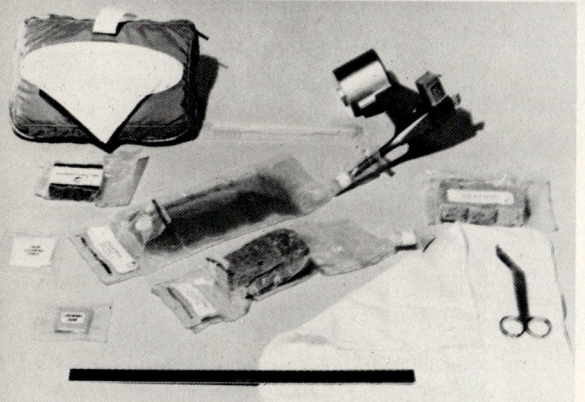

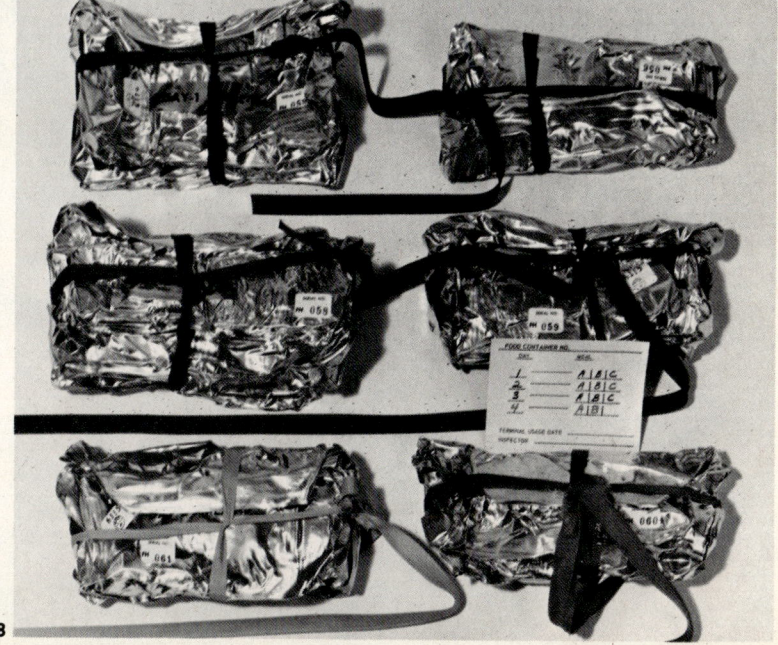

THE RED, WHITE & BLUE

The new color-coded space meals may be a far cry from the astronauts' steak-and-eggs earth breakfast but are equally nutritious, nevertheless. Gone are the days of crumbling bite-size cubes for Scott Carpenter, or the obstinate package rehydration reported by Gordon Cooper. Although the act of eating in a zero-gravity situation is a new experience and not comparable to eating on earth, and although food must withstand decompression and other space rigors, the fact remains that NASA has produced startling results in spacefood development. On earth, the astronauts begin a blast-off day with a thick, tender cut of beef steak, along with scrambled eggs and coffee. NASA requires it. Once in space, a typical Apollo meal will include shrimp cocktail, chicken and vegetables, toast squares, butterscotch pudding, and apple juice. Because the menu is varied, astronauts face the repeat of a particular meal only about every fourth day. With the advent of the Apollo 8 Christmas flight, astronauts Borman, Lovell and

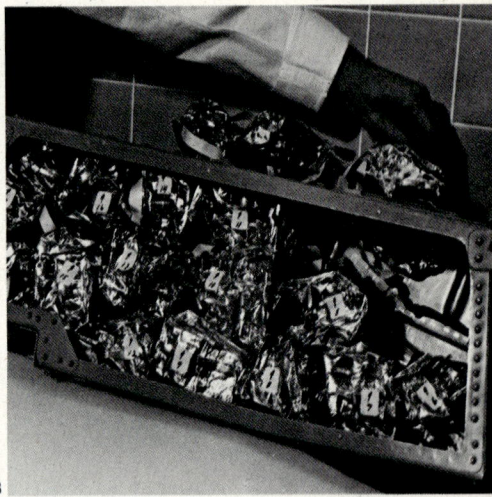

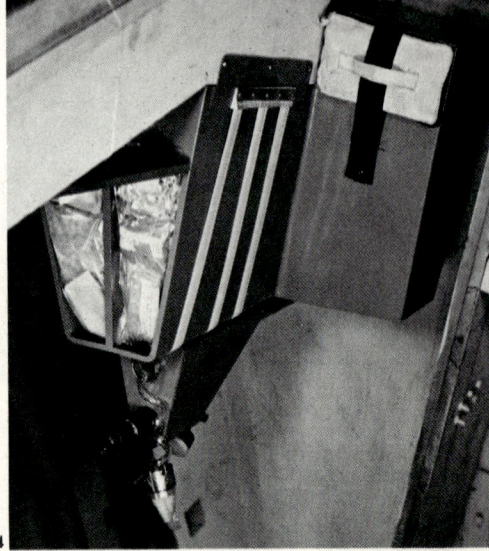

Anders demonstrated that space food could be eaten in a more natural way—with a spoon—as they opened their surprise packages of gravy-covered chunks of turkey with cranberry applesauce. One of the favorites is this bite-size sandwich (**1**) which an astronaut can pop it into his mouth easily. But a semi-liquid food must be extracted with more care through a plastic tube (**2**) following rehydration. If the container is not properly sealed, the liquid will float out of the tube. But color-coded food doesn't present a problem, with individual arrow marking and a tightly packed container (**3**). An earlier Gemini mock-up featured integral bulkhead food lockers (**4**). And, to capitalize on almost Lilliputian storage capacities, NASA devised special kitchens (**5**), with sensitive scales (on counter) to assure proper weight and consistency. Not neglected is continuing basic research, as this technician experiments with a rehydrated cheese product (**6**). The bite-size sandwiches are cut from regular sized favorites in NASA kitchens (**7**), at the Manned Spacecraft Center. What a wonderful invention... the sandwich. The most important device for man's culinary attack on outer space, however, seems to be the water dispenser (**8**). Without it, the Apollo program would be back to the unappetizing days of toothpaste tubes. Future space flights are dependent on space food evolution. It is a science still too young to change eating habits on earth, but because several commercial products are related to NASA developments, perhaps more will follow.

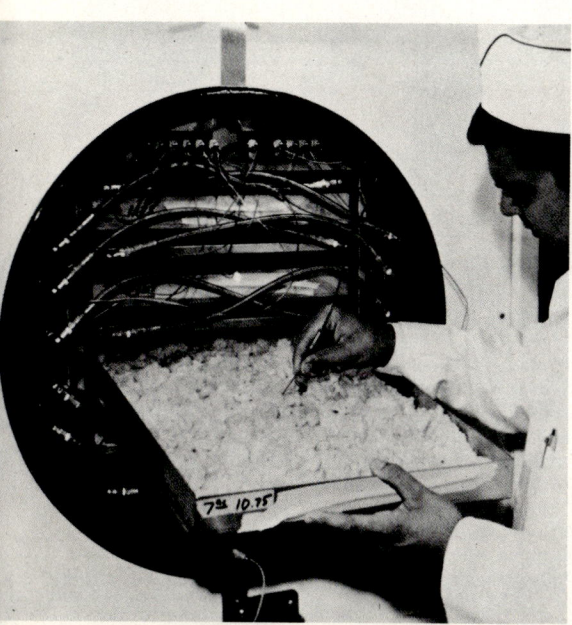

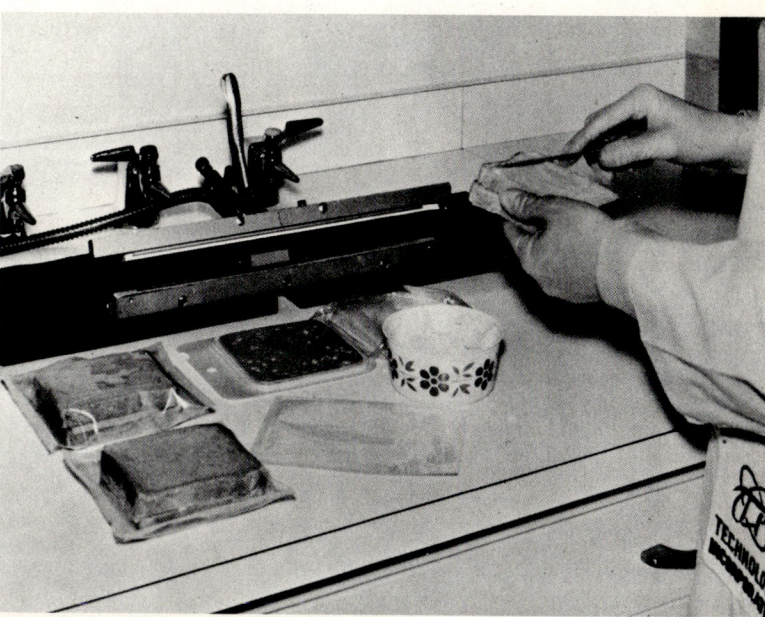

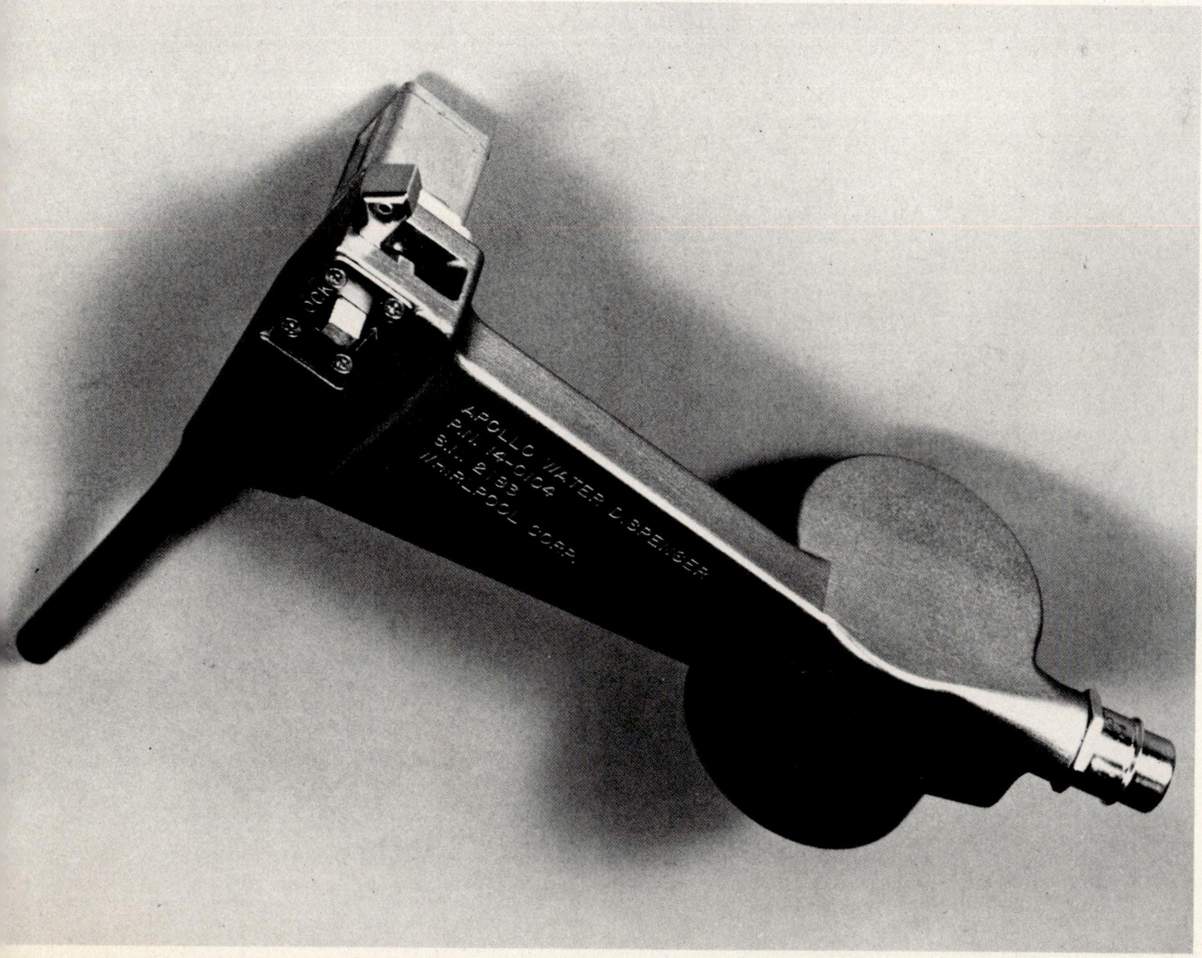

APOLLO 8

There will be a difference in this winter's night sky, when all is cold and still and the moon shines across the earth in wide swaths of silver. The difference will come with man's first orbit of another celestial body, this very same moon. Apollo 8 will be a Christmas journey, and the birth of manned interplanetary flight is as magnificent a Christmas gift as the world could wish. This will be the first manned flight of NASA's new Saturn V three-stage booster, and enormous pressure is now being placed on men and machinery to function perfectly. Once the third stage S-IVB engine fires, pushing Apollo 8 out of earth's gravitational sphere of influence, there will be no turning back. In the past, with earth orbits, re-entry could be accomplished quickly if anything went wrong. Now, a safe return to earth will be measured in days. Once at the moon, the crew will brake by retrofiring the Service Module's engine and enter lunar orbit. After ten photo-mapping circuits, they will restart the SM engine to return home. For this flight, NASA has selected an experienced trio, two of whom, Borman and Lovell, have logged more space hours than the entire Russian cosmonaut corps. Apollo 8's patch features a red "8", crisscrossing the earth and the moon.

FRANK BORMAN

SPACECRAFT COMMANDER/ Born in Gary, Indiana, on March 14, 1928, Borman is a Colonel in the USAF. He has a BS degree from the U.S. Military Academy at West Point, and an MS in Aeronautics from Cal Tech. A former fighter pilot and test pilot, he was the command pilot on Gemini 7, which performed the first U.S. space rendezvous and is still the longest space flight on record. He was a member of the Apollo 1 fire investigation team, and program manager for the resultant changes instituted in the Apollo spacecraft.

JAMES A. LOVELL, JR.

COMMAND MODULE PILOT/ Born on March 25, 1928, in Cleveland, Ohio, Lovell is a Navy Captain. He is a graduate of the Naval Academy, and a former test pilot at the Naval Air Test Center, Patuxent River, Maryland. He teamed with Borman on the historic Gemini 7 rendezvous flight, and was command pilot on Gemini 12. With a combined total of 572 hours from these two flights, he has logged more time in space than anyone else. Since June of 1967, he has been a President's Consultant on physical fitness and sports.

WILLIAM A. ANDERS

LUNAR MODULE PILOT/ Bill Anders, a Lieutenant Colonel in the USAF, was born in Hong Kong on October 17, 1933. He has a BS degree from the U.S. Naval Academy, and an MS in Nuclear Engineering from the Air Force Institute of Technology. An instructor pilot with the Weapons Laboratory at Kirtland AFB, New Mexico, he has logged over 3000 hours flying time and was selected in the third group of astronauts by NASA in 1963. Though a space rookie, Anders was backup pilot for the Gemini 11 mission.

APOLLO 8 Preparation

It could be said that preflight for Apollo 8 began eight years prior to launch. It was an operation that has seen thousands of separate components find their way to assembly and test points. Modifications that were incorporated as the result of earlier missions will cause delays in some areas, and design changes resulting from static tests often threaten even the long-range schedule. The task of coordination is formidable, as is the feeling of relief and accomplishment when a major component finally arrives at the Cape (**1**). The second stage of the Saturn has just arrived by barge from Alabama. Meanwhile, in the Launch Assembly Building, the first stage (**2, 3**) of the Saturn V is being lifted off the surface transporter. Note that the fins and shrouds have been removed for ease of transport; a clear view of the gimbal motors and their bracing is possible. Once the S-IC stage has been checked out, the S-II or second stage is hoisted into position (**5**). The entire assembly of the Saturn V takes nearly a year, from arrival to actual launch.

During the assembly process, the entire rocket is erected on the Crawler/Transporter inside the Vehicle Assembly Building (VAB). The building represents the high point in functional design, with the largest volume of interior space of any building in the world. Without the VAB, the space program would be nearly impossible. More probably, reliance would still be on the smaller, pre-Saturn vehicles, with assembly of the lunar craft in earth orbit. The decision to embark on a direct earth-to-moon launch was decided in 1962, with construction of the launch complex begun shortly thereafter. During the months of assembly and test, the great VAB doors remain closed. But at last, on Oct. 9, 1968, they open; and like an awakened God of the new mythology—science—Apollo 8 comes forth. In its coat of gleaming white paint (**7**), Apollo/Saturn 503 stands atop the largest land based moving object, towering 363 feet above the Crawler/Transporter. This latter (**6**) is a squat colossus which travels at a dignified one mile per hour when loaded, on its 3½-mile journey to Launch Complex 39A. There meticulous "grooming" will take nearly two months, especially of the Command Module (CM #103). During the period, astronaut training continues (**4**), with added emphasis on lunar topography. CM pilot James Lovell (center) points to a map of lunar terrain that he and mission commander Frank Borman (extreme right) are expected to photograph from their orbital path of some 70 miles altitude. LM pilot Anders is behind Lovell.

APOLLO 8
Pre-launch

The morning of December 21 begins at 2:36, according to schedule. Breakfast is followed by a preflight physical and the suiting procedure. Technicians descend *en masse* upon the astronauts, with not much time for humor. But when a technician presents Frank Borman with a Christmas stocking (**1**), humor begins cropping up in small ways. The stocking serves as a reminder that in spite of a highly technical space program, it is man who is the focal point. The launch day has been preceded by an intense training schedule for the three astronauts and their back-up crew. William Anders, the Lunar Module pilot (**2**), here prepares for a simulated flight in the altitude chamber. And the prime crew of Apollo 8 spent long hours practicing (**3**) in the mission simulator. After suiting-up, the three astronauts walk the short distance to the van that will take them the 8 miles to the launch pad (**4**). Commander Frank Borman waves to a well-wisher in the pre-dawn darkness. CM pilot James Lovell is in the center, fol-

lowed by Lunar Module pilot William Anders. All are carrying the portable air-conditioners. Meanwhile, within the Launch Control Center, engineers monitor pre-launch activities for the Apollo mission (**5**). Actually, the countdown began on Dec. 15, 1968, at 7 PM EST when electrical power first flowed into Saturn A/S 503. Now there are about 200 persons at their places in the firing room, all under the command of Launch Operations Director Rocco Petrone. Fueling has been completed—with almost a million gallons of liquid oxygen, liquid hydrogen and other propellants pumped aboard. The astronauts enter the Command Module, and the hatch is closed at 5:34 AM EST. The final checklists are being re-checked as the "GO" comes in from various sections of the massive operation—places like weather, range tracking network, and all the parts of the intricate coordinates which now appear as green lights on the big board in the Control Center. Because there are no "holds," the countdown proceeds into the final minutes. Outside, the crowds wait. Some, like those at the jetties (**6**), just south of the Cape on the Atlantic side, have been waiting several days in campers and tents—to see the lift-off of the first manned vehicle to leave earth orbit. Others, with a more personal interest are located about one-half mile from the VIP site. Mrs. Marilyn Lovell (**7**) holds her youngest child, Jeffrey, with daughters Susan and Barbara. Also watching the fiery lift-off is Charles A. Lindbergh, the great hero from a bygone era of flight.

APOLLO 8
Launch

APOLLO 8
In-flight

orange flame, the crowd breaks into a single, drawn-out "Ohhh;" and then falls silent, as Apollo 8 lifts with agonizing slowness past the umbilical tower. The sound and shock thunders across the water (page 46-47), 15 seconds after lift-off. In the Launch Control Center, tension mounts until "tower cleared" is announced. At this point, control passes to Mission Control in Houston for the remainder of the long journey. The time is 7:51 EST, and the first manned Sat-

the astronauts glue their attention to the panel displays, a kaleidoscope of up-dates. There are 24 instruments, 71 lights, 40 event indicators and 560 switches. Borman has override capability in the event that the automatic sequencing and control system fails. But everything goes smoothly. The first stage S-IC separates, followed by Launch Escape Tower jettisoning. In succession, there is second stage burnout and separation; a brief coast, then ullage—a quick

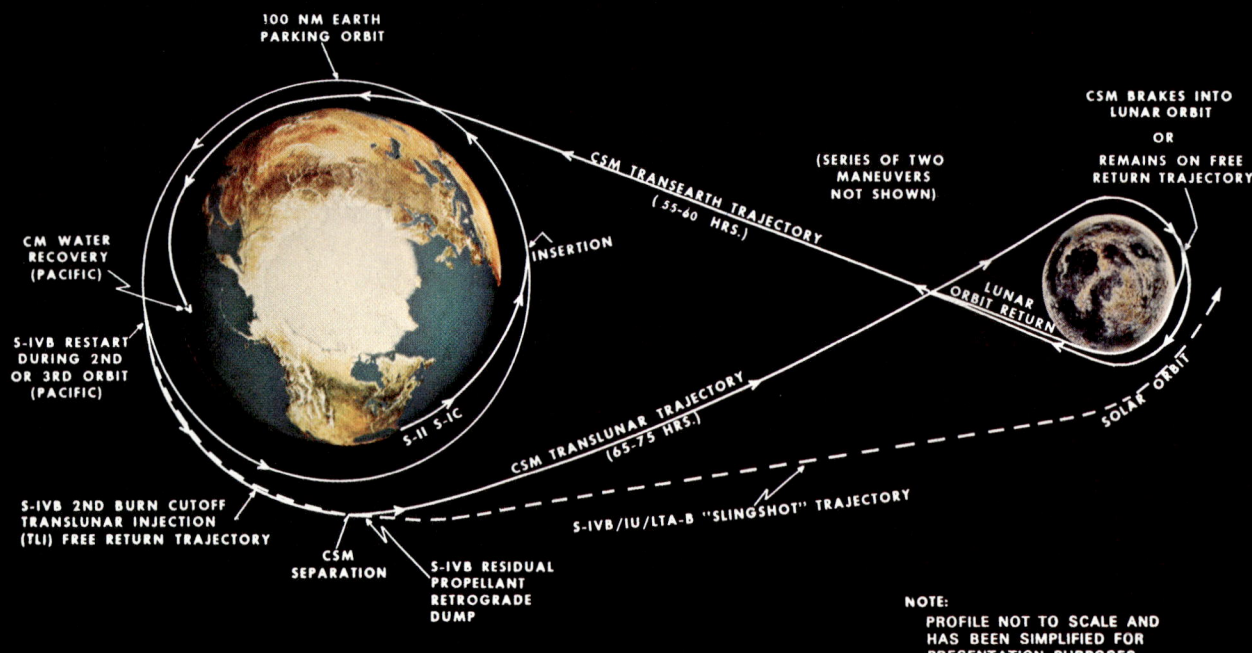

thrust of the vehicle, made before firing the single J-2 engine of the S-IVB third stage, to shift the propellant to the bottom of the tanks so it will feed properly. The third stage S-IVB fires until orbital velocity is obtained, 11 minutes after lift-off. The speed is now 17,428 miles per hour, in a circular 100-mile orbit. Lovell immediately unstraps from the center couch and begins his navigation duties. His fast movements, though, together with an unfamiliar sensation of not being strapped down (as he was on his two prior Gemini flights), cause him to become queasy. Lovell slows his movements, and the feeling disappears. All systems, however, are checked out. Once the translunar injection burn is made, there is no turning back. The word is "GO." The guidance system reference alignment has taken place, and the craft is oriented with the thrusters for the translunar burn. At 10:41 AM EST, the S-IVB begins the longest burn of the mission, lasting a little more than five minutes, and boosting the speed of of the spacecraft to 24,226 miles per hour (**1**). It is an almost perfect trajectory. Borman pulls the T-handle that triggers the explosive device severing the S-IVB from the CSM. It is the first manual control of the flight. The S-IVB stage (**2**) is, however, too close for comfort, and is spewing fuel from all sides. Lovell reports: "I am looking through the scanning telescope and I see millions of stars." They are fuel droplets, and seriously interfere with the star sightings. A short burn of the attitude control thrusters is required to get clear.

APOLLO 8
In-flight

After getting clear of the fuel droplets, the astronauts engage in some spectacular sight-seeing (**2** and **4**). As seen from 3500 miles out, the crew gaze upon the shallow Bahama Banks (**2**), and a short time later, Australia (**4**). Several small problems crop up, though. Anders reports that three of the five windows are fogged up. This is due to the sealing compound giving off gas between the triple layers of glass that make up the windows. On an unmanned flight, there would be no way to compensate for this; now, the astronauts simply shift their equipment to the two remaining windows and continue their tasks. But never before has man seen his own home from such a vantage point, 21,000 miles away (**6**). Dramatic changes are now taking place to the spacecraft, itself. Earth's gravity pull is a powerful force on Apollo 8's velocity. In just three hours, it has decreased from 24,200 to about 8,500 miles per hour. The distance from earth is now 26,000 miles. When the star sightings and picture taking are over,

the spacecraft is put into Passive Thermal Control. This is called the "barbecue mode" by the crew because, in it, the spacecraft revolves at one revolution per hour, allowing for an even heat exposure in the direct sunlight of space. A short mid-course correction burn of 2.4 seconds is made some 60,000 miles from earth and it adds an additional 17 miles per hour to the spacecraft's velocity. Commander Borman finally turns in for a needed rest. He has trouble getting to sleep, however, and has to take a sleeping pill. Shortly afterwards, he is beset with nausea, vomiting and diarrhea. It is not reported at the time and will only be discovered ten hours later during a routine tape play-back by Mission Control. Borman does manage to shake off the effects of the pill—or possibly the flu—while the other two crew members show no ill effects. The mission, therefore, proceeds as planned. At 3:01 PM EST on Dec. 22, the first telecast is made from the spacecraft at a distance of 120,653 nautical miles from earth (**1, 3 and 5**). The TV camera works perfectly. Borman (**1**) works the controls, Lovell (**3**) wishes his mother a "Happy Birthday," and Anders demonstrates his floating toothbrush (**5**). Interestingly, both the earth and the moon are too bright for any visible detail at this distance. On Dec. 24, Apollo 8 passes from the earth's to the moon's gravity. It happens at about 39,000 miles from the moon; and the spacecraft's speed, which has been decelerating to a low point of 2,220 miles per hour, will now begin a gradual climb to over 5000 mph.

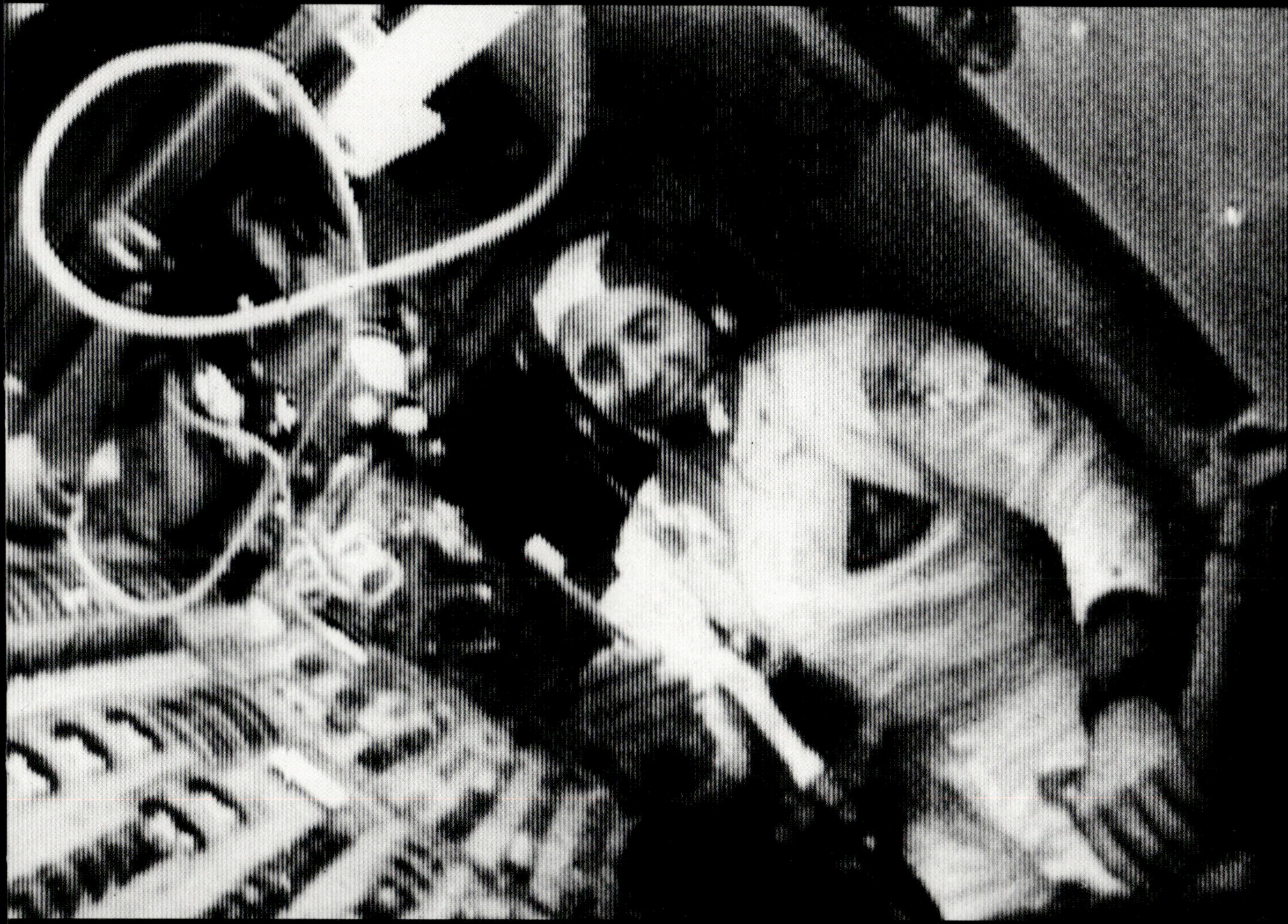

APOLLO 8
In-flight

The busy crew schedule includes periodic photography of both the earth and moon. Most of it is done through the telescope; and various filters (**2, 5**) are used. Some photographs reveal features that had not appeared on the Lunar Orbiter mission or the huge earth reflector telescopes. The bright rays, which radiate from a large crater (**1**) at the top left-hand side were not visible in photographs taken by the unmanned Lunar Orbiter space probes, due to low sun elevations. Actually, the lunar surface has considerably less color than is shown, and the crew of Apollo 8 sees very little of the moon on the journey from earth. Borman remarks that "It's like being on the inside of a submarine." This is due to fogged windows resulting from the barbecue mode. Lovell, who operates the telescope at his station, has the best view. But now, Mission Control is very busy cross-checking data for the Lunar Orbit Insertion burn. This is the critical retrofire burn which will place Apollo 8 in lunar orbit, and it will take place with the

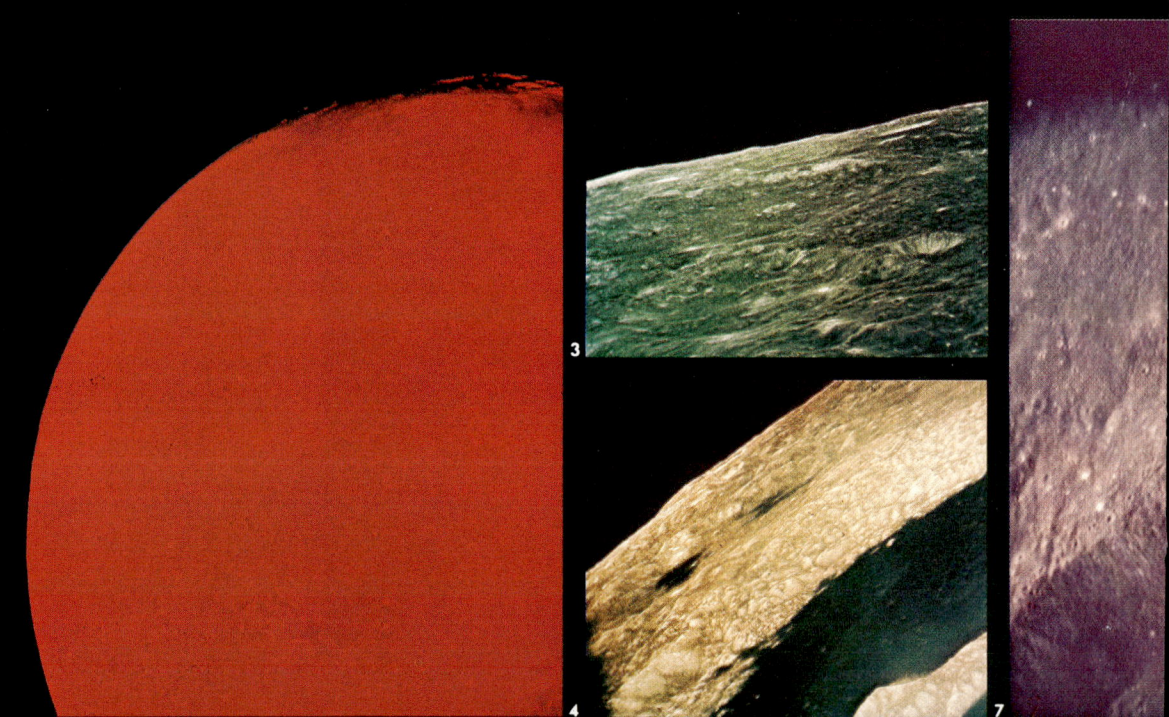

spacecraft on the far side of the moon, out of radio communication with earth. A slight reduction in forward velocity is needed. The small CM reaction jets are fired for 12 seconds, slowing the spacecraft by 1.4 feet per second. At 4:49 AM, Dec. 24, radio communication with Apollo 8 is lost; an anticipated but tense waiting period begins. Apollo 8 is due to reappear on the other side of the moon in 34 minutes, and it will be this long before anyone in Mission Control knows what has happened. This is the first time that communication has been lost for an extended period. During re-entry to earth communication is lost for several minutes, but this happens after all the critical maneuvering and retro-firing has been accomplished. But all goes well. In a four-minute burn of the Service Module's SPS engine, man is placed, for the first time, in orbit around a heavenly body other than his own planet. At this moment, the astronauts see what no men have ever seen—the far side of the moon (**3, 4, 6, and 7**). Following 35 minutes after loss of radio signal, telemetry is again established, and Lovell gives the first verbal description: the moon is. . .''essentially grey; no color, looks like plaster of Paris, sort of greyish sand.'' Anders, who is the cameraman, reports that the moon is ''whitish grey, like dirty beach sand with lots of footprints in it.'' The crew recognizes most lunar features, including the large crater Langrenus (**9**). In awed silence, they become the first humans to witness a stunning spectacle of space (**8**): earthrise over another world.

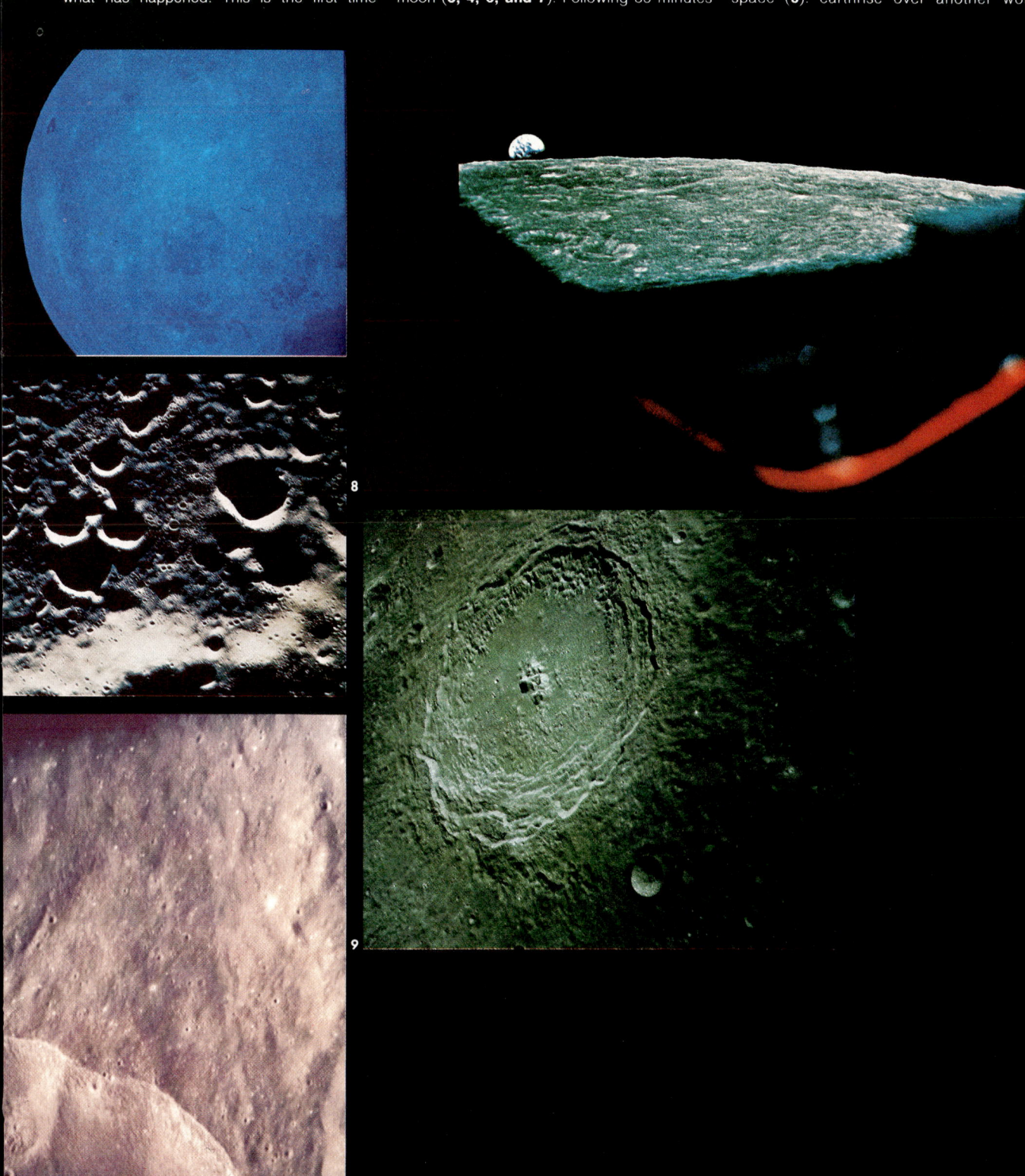

APOLLO 8
In-flight

watching the vast, dark emptiness of space and the lifeless expanse of lunar surface, Borman, Lovell, and Anders are very moved. As the stark, barren surface of the moon moves across the TV screens of the world (**3**), Anders begins reading from the book of Genesis..."In the beginning, God created the heavens and the earth. And the earth was without form, and void...and darkness was upon the face of the deep." They are words from an age of long ago, when man trembled at many things and was comforted by the thought of an ever-present God. In a way, Apollo 8 is like a journey to God's workshop, a primeval world untouched and even unseen by men until now. And so the thoughts go. For after all, it is Christmas Eve on earth. The crew opens a special dinner treat—a foil-packaged portion of turkey and cranberry sauce. While the astronauts eat, a listening, awe-struck earth is suddenly aware that these three men are very far away, indeed...and if they should need help, we view them perhaps as sailors of legend, who ven-

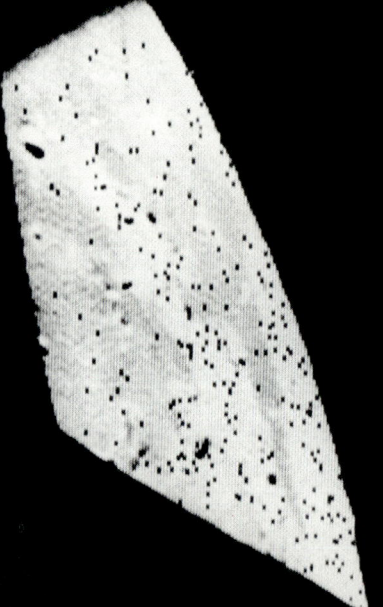

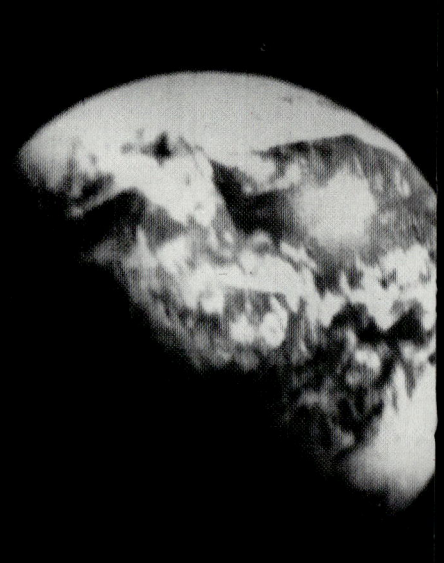

was thought flat. Within the spacecraft, the effort of keeping up with the crowded flight plan is beginning to tell. Lovell has made an error on the computer phase, causing the computer to lose part of its memory. Mission Control has to spend several hours in checking that the critical re-entry portion of the memory bank has not been erased. With the Transearth Injection (TEI) burn only hours away, Borman decides to cancel the remaining experiments. His crew badly

the work load has been too heavy, and the crew schedule too tight. Fatigued men make mistakes, and in space, even a small mistake can spell disaster. It is remembered that Scott Carpenter, in *Aurora 7*, missed his splashdown point by 250 miles because he was only 5 seconds too slow in retrofire. Several hours before midnight on Dec.24, preparations begin to leave the moon. Again the burn takes place behind the moon, away from earth communication. The service pro-

during the 10th lunar orbit, accelerating Apollo 8 out of orbit. Afterwards, the crew relaxes, eats, and sips small airline-type bottles of brandy. Borman (**5**), Anders (**1**) and Lovell (**2**) take pictures; and during the final telecast, Dec.26, Anders says: "I think must have the feeling that the travelers in the old sailing ships used to have. I have the feeling of being proud of the trip, but am still happy to be going home... We'll see you back on the good earth (**4**) very soon."

APOLLO 8
Splashdown

So far, the trip home has been uneventful. But now, 30 minutes before splashdown, there is a burst of activity. Cameras, checklists, maps and other pieces of equipment must be stowed and secured before reentry. At nearly seven-G's deceleration, a 1-pound camera that breaks loose will come hurtling down with a force of 7 pounds, enough to injure men and equipment. Now, the Service Module is jettisoned, leaving the Command Module on its own supply of power and oxygen as it rapidly approaches earth's atmosphere. The capsule is traveling at some 25,000 miles per hour; fast enough, if the angle of flight into the earth's atmosphere is too steep, to burn the spacecraft to a cinder; or, if the angle is too shallow, to bounce it back into space. To splashdown safely, the capsule has to be threaded through a space "window," that in reality amounts to the eye of a needle—taken in proper perspective at 400 miles high by 26 miles wide. At 400,000 feet, the astronauts feel the first hint of deceleration caused by

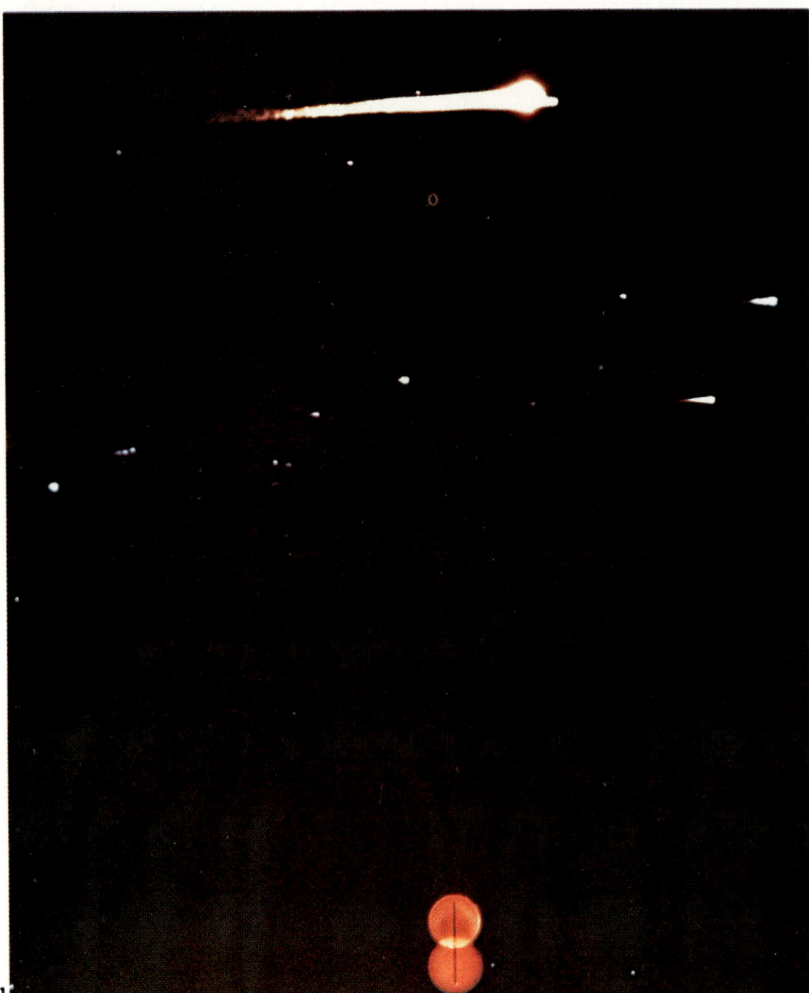

atmospheric friction; and radio blackout begins. Twice the craft is rolled so that the aerodynamic lift designed into it not only slows descent, but actually causes a brief climb. For the first time the blazing re-entry is captured on film (**1**) taken by an Airborne Lightweight Optical Tracking System (ALOTS). The camera is mounted on a KC-135A jet tanker aircraft at 40,000 feet above the Pacific. Three drogue parachutes deploy automatically at 24,000 feet and Apollo's speed slows to about 300 miles per hour. By 10,000 feet, the capsule is slowed to some 140 miles per hour, and the three main 83½-foot parachutes are pulled free into deployed position. Dropping like an autumn leaf, the capsule splashes into the Pacific waters at an easy 19 miles per hour. It is 10:51 AM EST Dec. 27. But 1,100 miles southwest of Hawaii, it is 4:51 AM. One hour and 20 minutes later, the crew steps safely aboard the recovery carrier USS *Yorktown* (**2**), and the return is celebrated at Mission Control (**3**). The next day, there is a welcome in Hawaii (**4**), given in the delightful native manner. Later, an address is given at the Smithsonian Institution (**5**), under two of the most historic aircraft in the world, the Wright Brothers biplane and Lindbergh's *Spirit of St. Louis.* Governors and mayors are delighted to be presented to the astronauts (**7**); shown is Governor John Connally and Houston Mayor Louis Welch. The climax comes at a meeting in the White House (**6**), as President Nixon announces a European goodwill tour by the famous crew.

When you consider the primary goal of our space program—landing a man on the moon before the end of the 1960's—then the Lunar Module becomes one of the most critical pieces of hardware in that effort. But when you realize that reaching the moon has been the dream of men for centuries, then the Lunar Module takes on an even more vivid role in history. This is the "rocket ship" that will do the job. The technology of our time has taken away the story-book cigar shape and fairy tale terminology of past dreamers, but the aura of mystique is even more immense as a result: in our time, it's no longer a fairy tale. As long ago as 1923, the German researcher Hermann Oberth postulated that a spacecraft for interplanetary exploration would carry its primary fuel supply in a "capsule" which would remain in orbit while the manned portion would descend to the surface. Upon return from the surface, the manned stage would reconnect with the fuel capsule and return to earth. Thus, Oberth, if not the first, was the first in

THE LUNAR MODULE
An ugly duckling designed for use only on other worlds

modern times to describe the elements of our present state of orbital science. For a long time NASA was of the opinion that the best way to send a man to the moon was by a direct shot there and back. In other words, the entire vehicle would be streamlined and capable of re-entering earth's atmosphere and landing intact. As time passed, however, it became evident that the best way to accomplish a lunar mission would be similar to Oberth's plan: an orbital vehicle, which came to be called the Command/Service Module, and a separate lunar lander that was designed to operate only in space, the Lunar Module (LM). Such a mission profile was delivered in a paper by H.E. Ross to the British Interplanetary Society on November 13, 1948. A mockup of an early design (**2**) was placed on display for President Kennedy at the Manned Spacecraft Center for his visit there on September 12, 1962. The design already incorporated a descent and ascent stage and had a modicum of streamlining, but the crew compartment would be changed. Grumman's model (**3**), built in 1962, contains all the elements of the final design; the mouth-like aperture is the exit hatch. The early testing of models (**4**) shows this development of the docking technique between a CM and an LM. As the design studies reach conclusion, Grumman technicians assemble an advanced mockup (**5**), which clearly shows the upper and lower parts, or stages. The artist's rendering (**1**) depicts the LM interior in its near-final form, with the two-man crew landing on the moon.

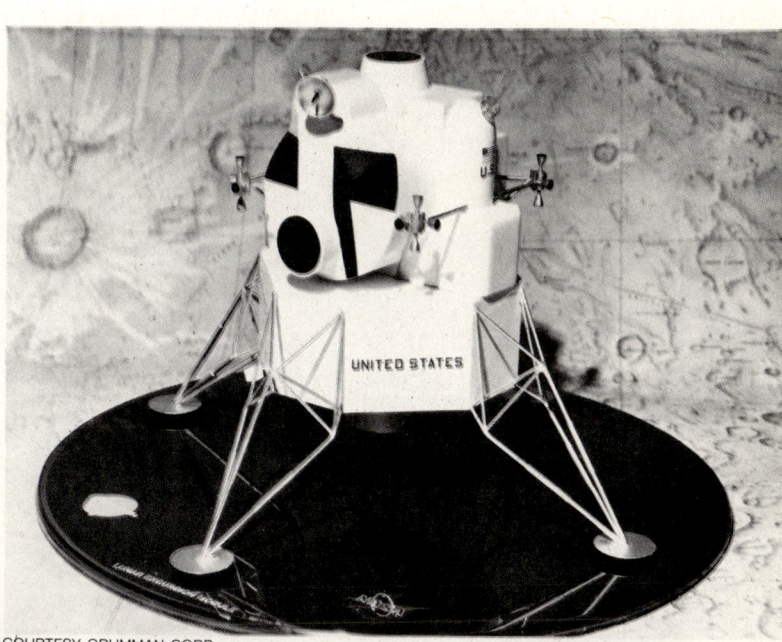

COURTESY GRUMMAN CORP.

LUNAR MODULE

It was not until July 11, 1962, that NASA officially chose the "lunar orbit rendezvous" method for the Apollo flights. Two weeks later NASA invited eleven companies to submit proposals for the Lunar Module, and on November 7 selected Grumman Aircraft Engineering Corp. as the winner. Grumman, located in Bethpage, Long Island, is the aircraft firm famous for its rugged Navy fighter planes. The initial contract was for $350 million, to produce 15 flight vehicles, 10 test vehicles, and 2 simulators. The two-stage concept was decided upon early. Since the LM will land on the moon only once, the complicated landing mechanism is not needed once the LM leaves the lunar surface. It is therefore left behind in the *descent stage*, which includes the variable-thrust landing engine and its associated tankage. The upper half, or *ascent stage*, includes the crew quarters and control area, all necessary life support and guidance equipment, and the ascent engine. When the time comes to leave the moon, the crew detonates explo-

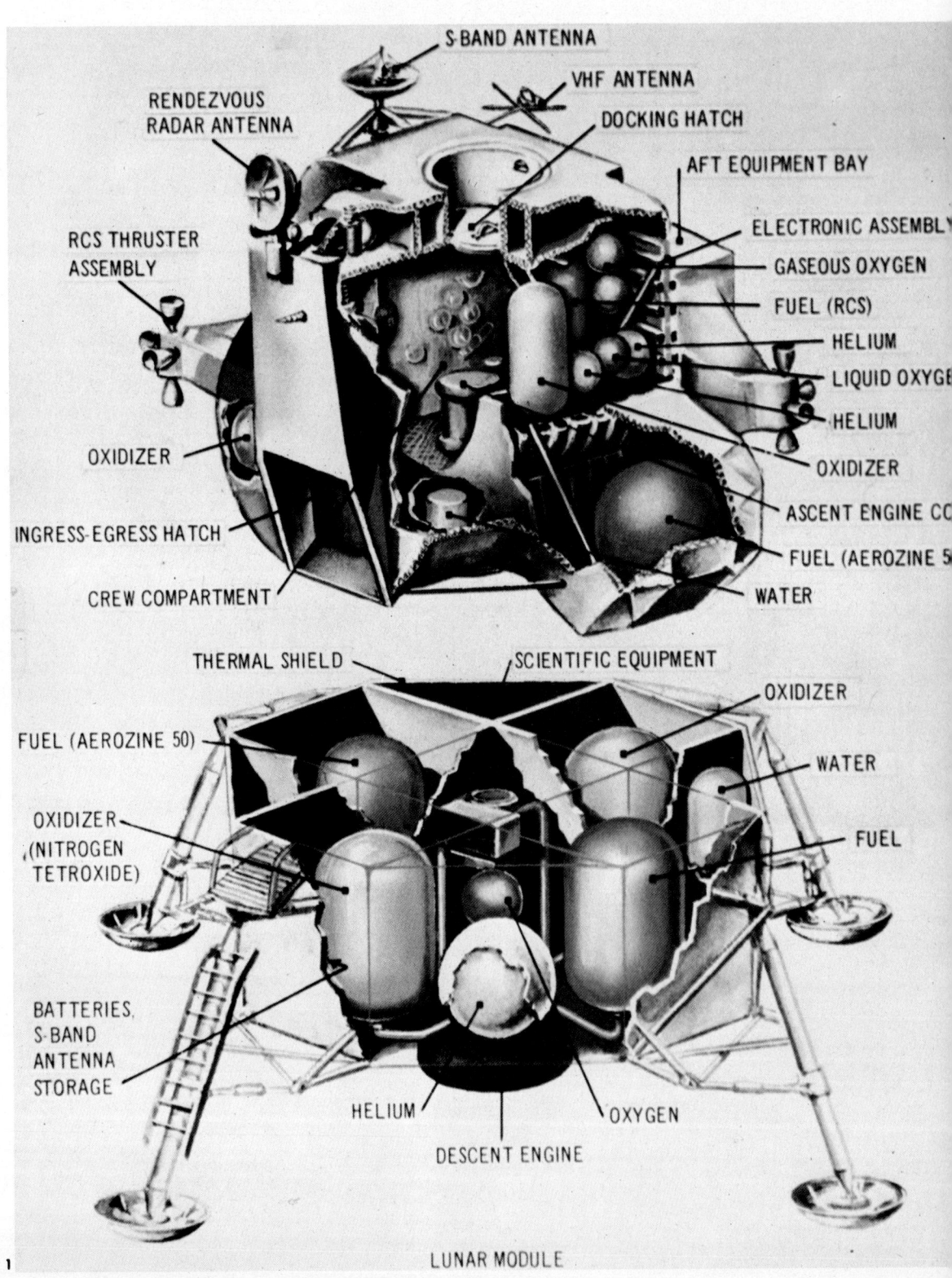

LUNAR MODULE

sive bolts which hold the stages together, and little guillotine-like blades chop through all electrical and hydraulic connections between them, leaving the ascent stage free to lift off. A NASA rendering (**1**) shows the two stages and their equipment. This view of the basic structure of the ascent stage during manufacture (**2**) shows the vertical "cold rails," through which passes coolant which maintains a safe temperature for electronic gear. Gas and liquid balance piping is here being fitted (**3**) at the bottom of the descent stage. In the center, the square cavity will soon be occupied by the descent propulsion system (DPS) engine. Now in place (**4**), the DPS engine has some of its controls wired near the nozzle skirt extension. This thin skirt deflects the rocket blast away from other components; the actual nozzle itself is high up inside. If the terrain at the lunar landing site is rough, this skirt extension can collapse up to 28 inches without damaging anything else. Built by TRW, the DPS engine is the only rocket motor in the whole Saturn/Apollo vehicle with the ability to "throttle"; its thrust can be varied from 1280 pounds to 9900 pounds. It can also be restarted 20 times, and has a total operating life of 910 seconds. Mounted on gimbals, it can be moved up to six degrees in any direction. This combination of steering and throttling ability is what allows the LM to maneuver during lunar descent. A full-scale mockup (**5**) shows the LM's size relative to the astronaut on the ladder; it is 22 feet 11 inches tall, and 31 feet across the legs.

COURTESY GRUMMAN CORP.

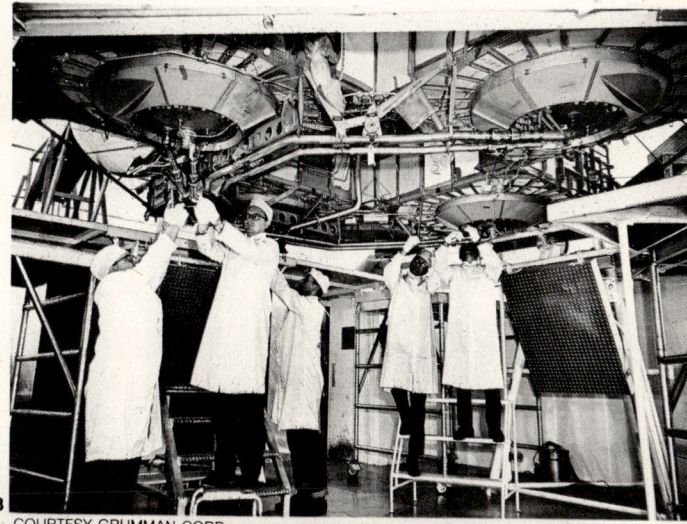
COURTESY GRUMMAN CORP.

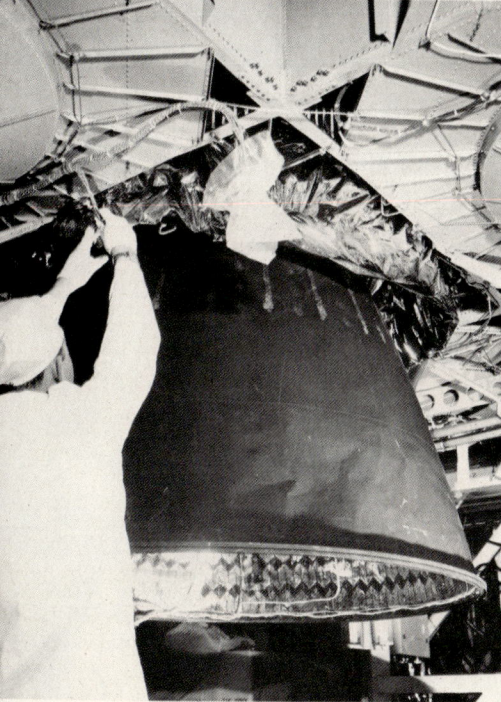
COURTESY GRUMMAN CORP.

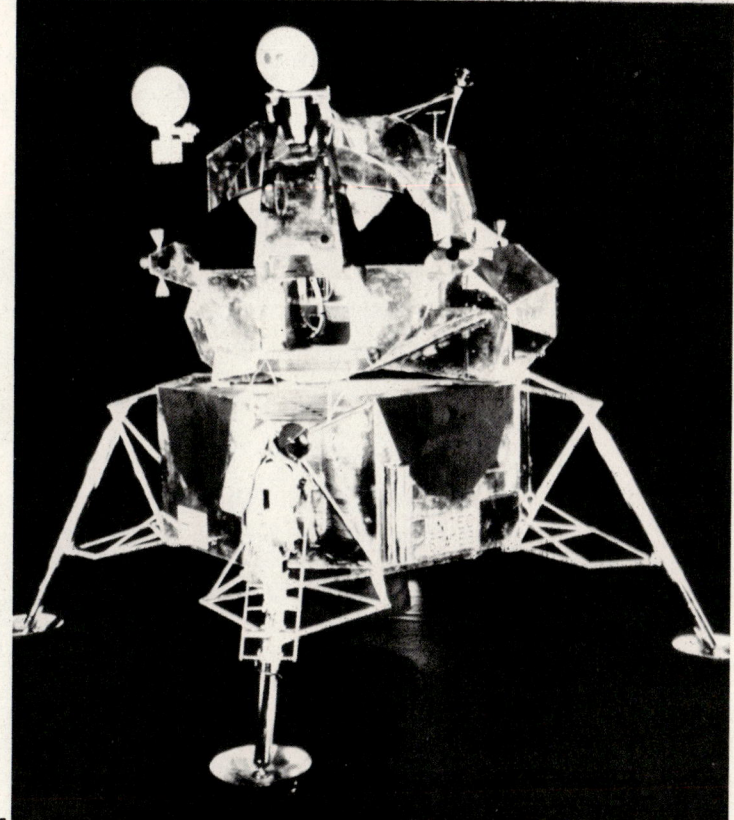
COURTESY GRUMMAN CORP.

LUNAR MODULE

Assemblers (**1**) complete the basic structure of the midsection of LM #9. Part of the ascent stage, this section lies directly behind the astronauts' flight stations; at the top is the docking tunnel. Both welding and mechanical fastening is used in construction of the LM, most of which is aluminum alloy. The midsection and crew cab are joined together here (**2**). These sections are carefully sealed, forming the cabin pressure shell. At the upper right can be seen one of the triangular windows of the crew cabin. Beneath it, in the projecting area of the front face, is the exit hatch through which the astronauts will have access to the lunar surface. After assembly, the ascent stage undergoes many checks, such as this test (**3**) of the environmental system. The completed ascent stage (**4**) wrapped and nearly ready for shipping, undergoes a last check at Grumman before being sent to Cape Kennedy; the covered nozzle of the ascent engine pokes out from the bottom of the stage. This engine is a combination effort of

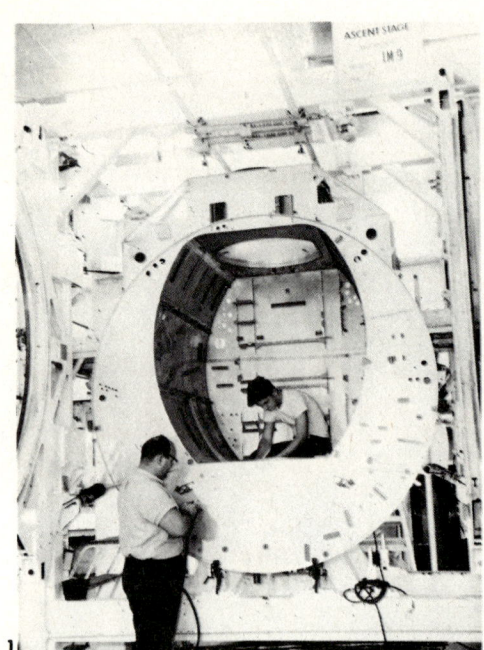
COURTESY GRUMMAN CORP.

COURTESY GRUMMAN CORP.

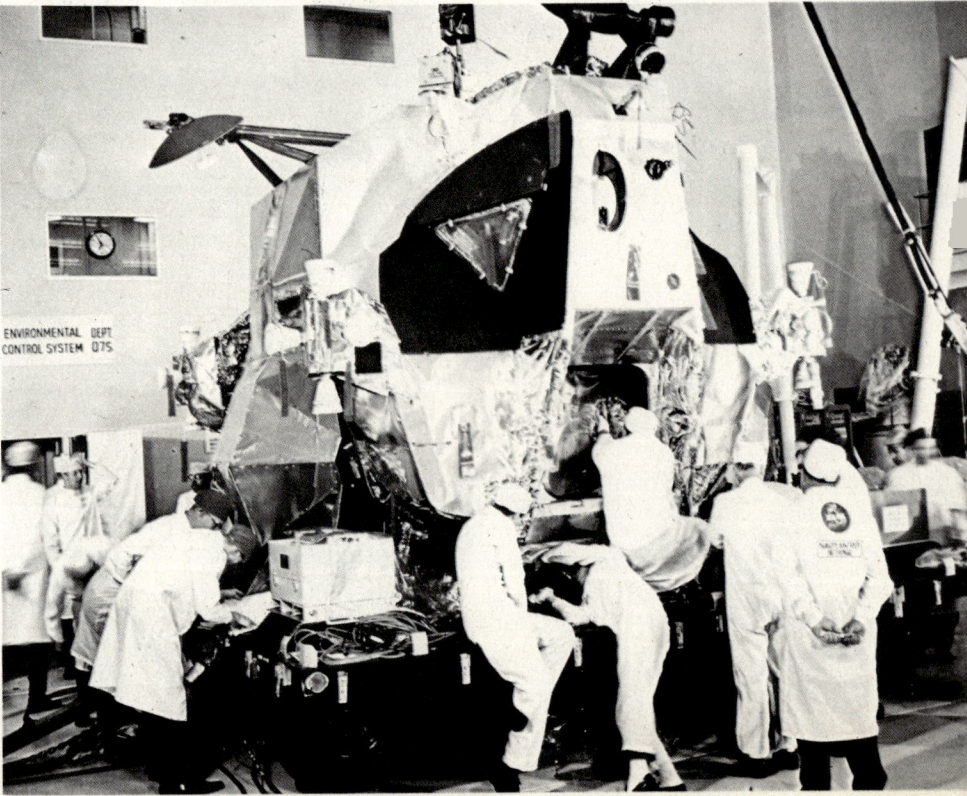
COURTESY GRUMMAN CORP.

two rocket specialists. The Rocketdyne Division of North American Rockwell builds the injector and combustion chamber, while the nozzle skirt and valves are made by Bell Aerosystems Co. This engine must have unfailing reliability, for without it the LM crew would be marooned forever on the moon. It is much simpler than the descent stage's DPS engine, being fixed (non-gimballed) to its frame and delivering a constant thrust of 3500 pounds. Both engines use the same propellants; the fuel is a mixture of normal hydrazine (chemical formula N_2H_4) and a variation known as unsymmetrical dimethyl hydrazine, while the oxidizer is nitrogen tetroxide (N_2O_4). This combination is "hypergolic," meaning that they ignite upon contact with each other, and therefore do not need an electrical igniter. Empty, the ascent stage weighs only 4700 pounds; it will be loaded with 5800 pounds of propellant, of which 600 pounds is for the small attitude control thrusters. The descent stage weighs 6100 pounds empty and takes 19,500 pounds of propellant. The complete and loaded LM thus weighs a little over 18 tons. Surrounding the ascent stage is the thin micrometeroid shielding, which absorbs tiny, high-speed space particles. The bright foil wrapped around much of the LM helps to maintain proper inside temperature levels. When finished, the separate stages are flown to the Cape in the oversize Super Guppy transport airplane (**5**). Each container is sealed and filled with non-flammable nitrogen, a final precaution against damage.

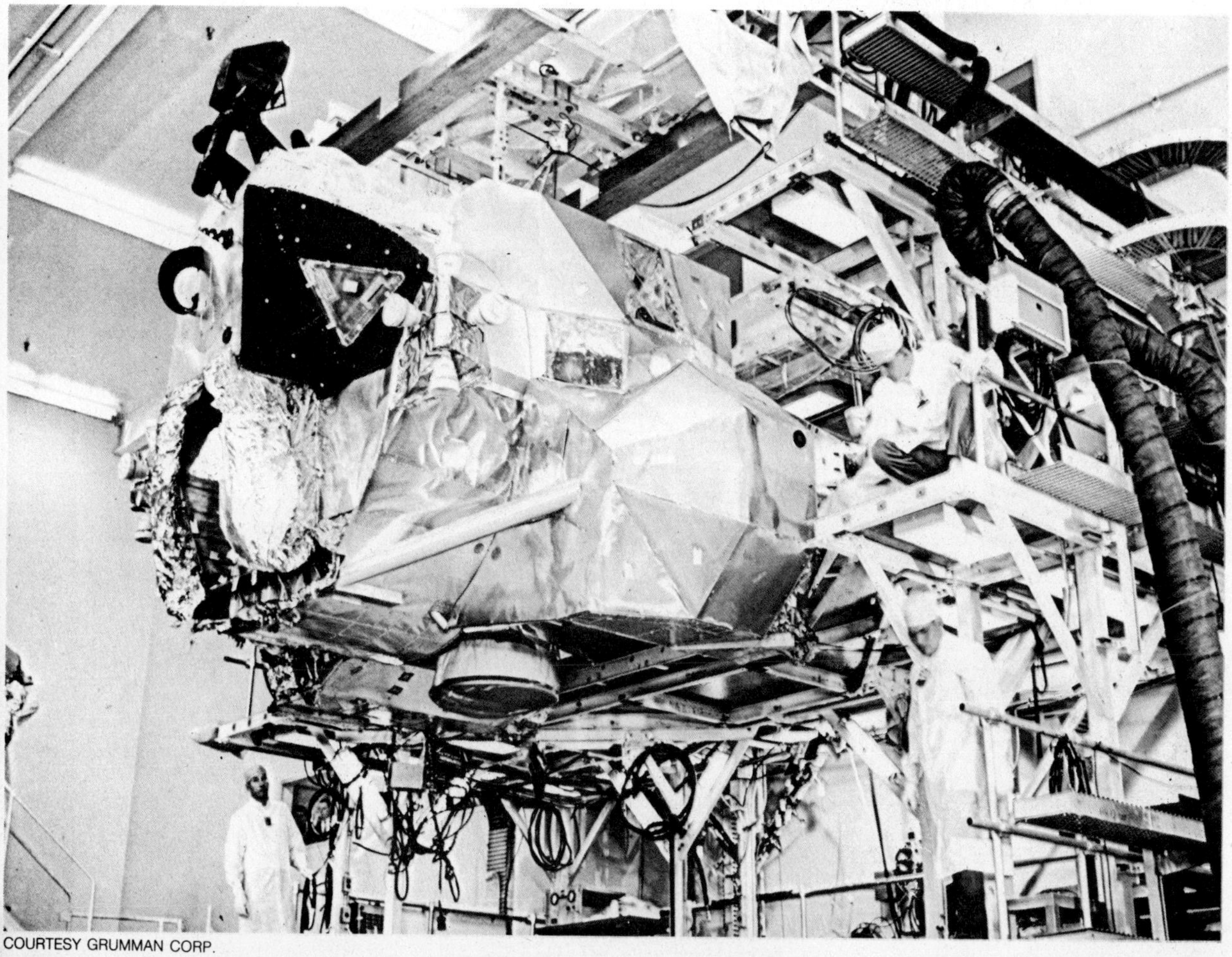

COURTESY GRUMMAN CORP.

COURTESY GRUMMAN CORP.

LUNAR MODULE

Training astronauts in the operation of the LM has not been the easiest task faced by NASA. Being experienced pilots, the crewmen can make the transition from aircraft controls to LM controls without great difficulty, but all training operations have to be carried out on earth. The landing on the moon will take place under different circumstances, the primary one being that the gravitational pull will be only one-sixth that of earth. To give astronauts practice in handling an LM, NASA developed this strange-looking flying machine (**1**), the Lunar Landing Training Vehicle; the astronauts quickly named it "the flying bedstead." A vertically-mounted turbojet engine in the center pushes upward with a thrust equal to five-sixths of the vehicle's weight, while the pilot maneuvers the craft by firing control rockets similar to the ones on the real LM. A sensitive device, it has proven susceptible to air currents. Astronaut Neil Armstrong almost lost his life when the machine went out of control during a test flight in May, 1968.

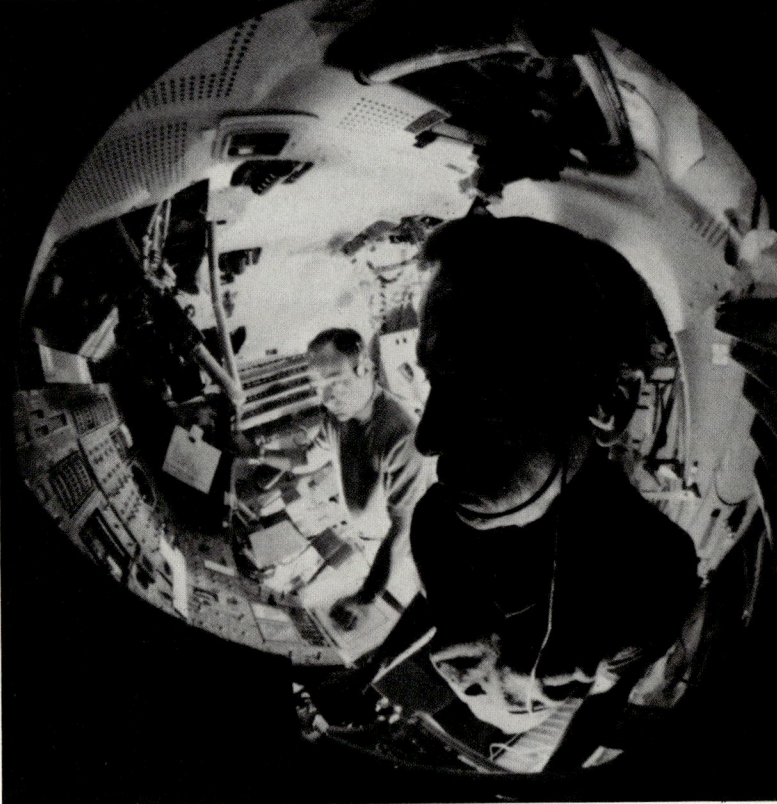

These remarkable photos (**2,3**) show his rocket-powered ejection from the tumbling craft, and then his safe descent by parachute, while the hapless LLTV burns fiercely in the background. At Langley Research Center, Hampton, Virginia, NASA engineers constructed this vehicle (**4**), another lunar landing simulator. In operation, it is very ingenious. A lone trainee stands in the cab, operating the rocket motor's controls. The vehicle is suspended from a crane and cables, part of which are visible. During operation, the cables apply a five-sixths lifting force to the vehicle's weight, but allow a range of flight travel of 360 feet down range, 180 feet vertically and 50 feet cross range. The arch-shaped suspension atop the vehicle permits six degrees of swing, duplicating the LM descent engine's gimballing action. A fisheye lens' view (**5**) shows astronauts Alan Shepard (foreground) and Edgar Mitchell in the Apollo Lunar Mission Simulator at the Cape. This sophisticated unit duplicates the LM's cabin, and is used to train for lunar touchdown as well as descent and ascent. Numerous emergencies can be "created" by operators outside the simulator. Films projected on the trainer's "windows" simulate a descent and landing on the moon. If a trainee "crashes," the windows turn red, providing a disquieting sensation of possible reality. This diagram (**6**) shows how the LM, returning from the moon, realigns with the CSM for docking. Optical sights with crosshair alignment are used, which center on opposing targets.

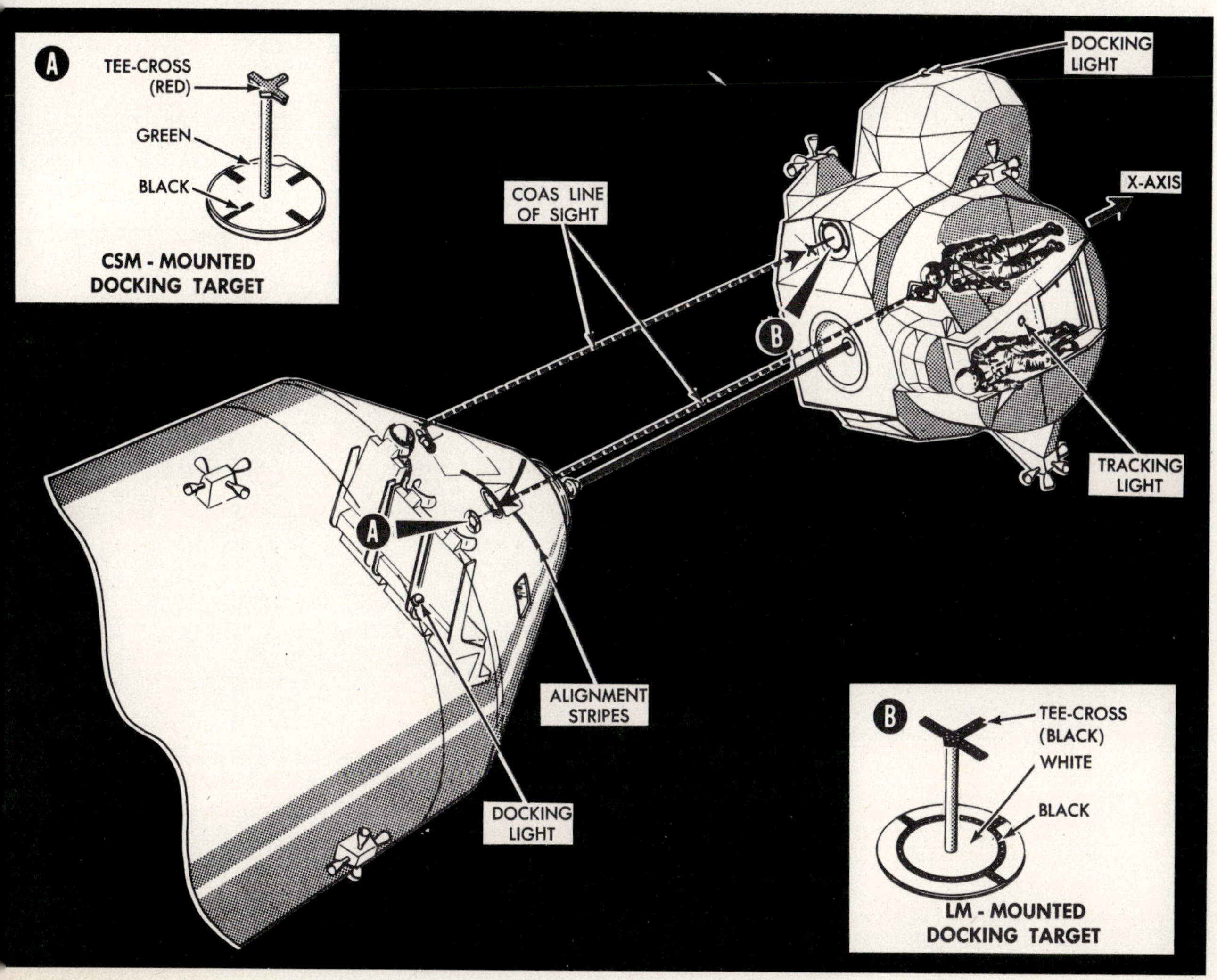

LUNAR MODULE

Within the ascent stage crew compartment is a complex but well-arranged array of vital instruments and storage spaces. The first diagram (**1**) shows the crew compartment interior, looking forward. In the center are the flight stations; the commander's is to the right, the LM pilot's to the left. Critical flight controls are either duplicated at both stations or accessible to both crewmen. Beneath Panel 4 is the forward hatch, the method of exit and entry when the LM is on the ground. Floor dimensions are not generous, measuring about 36 by 55 inches. The compartment is cylindrical, 92 inches in diameter and 42 inches deep; the whole crew compartment is scarcely larger than a good-sized closet. The second diagram (**2**) is the midsection, which is immediately aft of the crew compartment. The gray area shows the difference in diameter; there is an 18-inch step up from the crew section into this area, which is 54 inches deep and about five feet high. In the top center is the docking hatch, above which is the tunnel

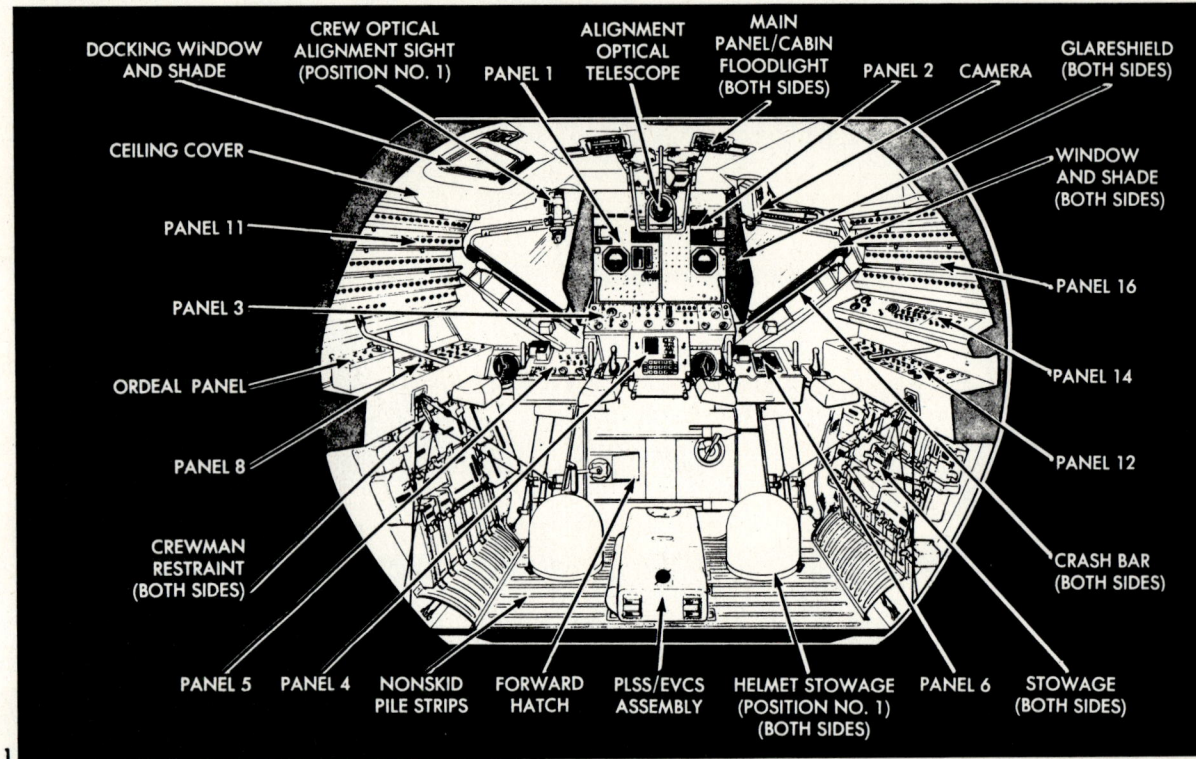

which connects with a similar tunnel in the Command Module when the two spacecraft are docked together in space. Important pieces of equipment are denoted. A considerable amount of space is devoted to stowage of equipment. Behind the midsection is the aft equipment bay (not shown). This unpressurized area houses electrical equipment as well as two oxygen and two helium tanks. No access is provided to this area during the mission. However, all the delicate equipment inside is provided with its own automatic temperature controls, as well as thermal and micrometeroid blanketing. The following two pictures (**3,4**) present an interesting comparison. To the left (**3**) is a mockup layout of the main control panels, made during the latter stages of LM design studies. A dummy Crew Optical Alignment Sight hangs from the overhead panel, while the two armrests protrude towards the camera in their unfolded position. The panels, with their control switches and gauges, underwent considerable revision from mockup to production, though the basic concept remained unchanged. To the right (**4**) is the fully operable panel of LM-6, the Lunar Module assigned to the Apollo 12 mission. The upper hatch (**5**) gives the crewmen access to the docking tunnel. Here the astronauts pass in and out of the LM from the CM. Approximately 33 inches in diameter, the hatch may be opened only after the cabin is depressurized. To the left of the docking tunnel hatch and below it (**6**) is the Portable Life Support System stowage area.

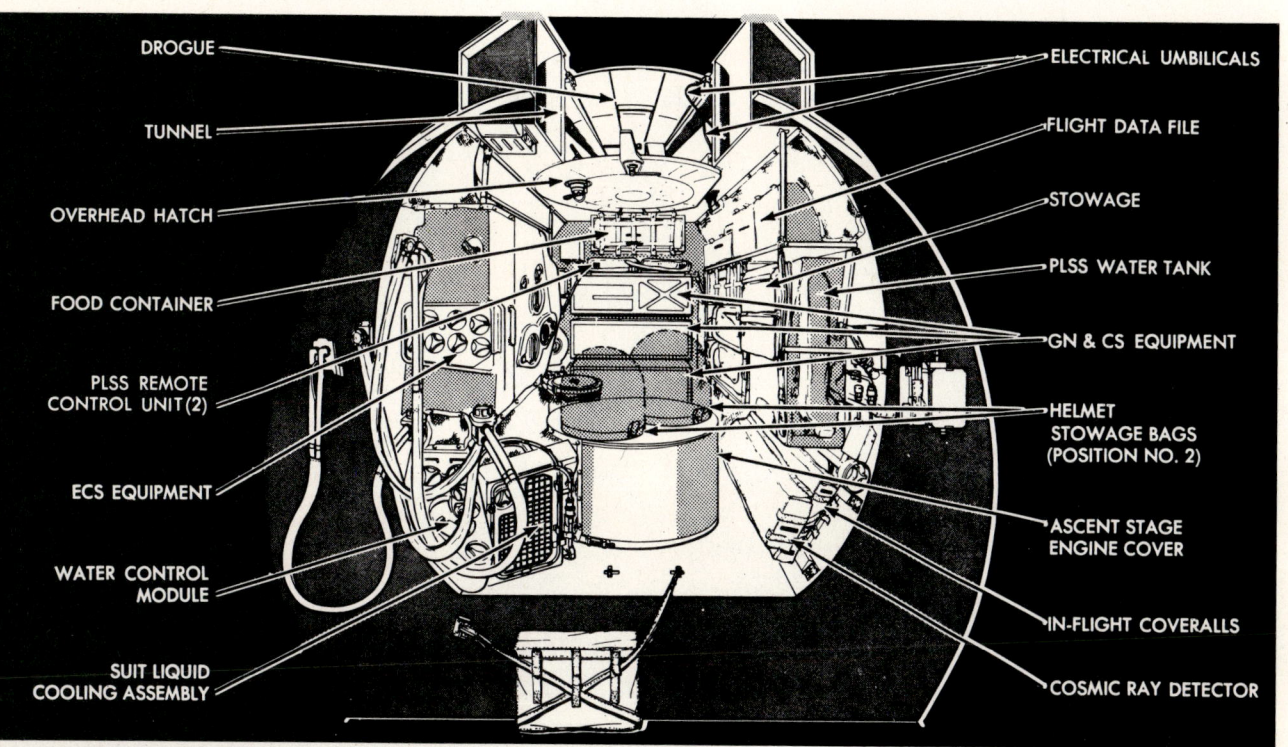

LUNAR MODULE

The camera looks downward (**1**) to the deck of the crew cabin on which are strips of non-skid material. The two EVA helmets and one of the PLSS backpacks are in their stowed location. The LM is no place for an astronaut with claustrophobia! Looking aft (**2**) towards the midsection, the Environmental Control Subsystem mounting point is seen. This unit provides pressurization of the cabin and suits, and regulates electrical system temperature as well as the oxygen supply of the cabin. It further provides drinking water, cooling water, and fire extinguishing. In this view, the LM is in shipping configuration and decked out with red "remove before flight" tags. Below the ECS pack is a large red plate for service personnel to use when climbing in and out through the docking hatchway. This is the view (**3**) seen looking forward from the midsection area, and the special lens makes the cabin appear much deeper than it really is. Astronauts Charles Conrad (left) and Alan Bean rehearse flight procedures; they are slated for

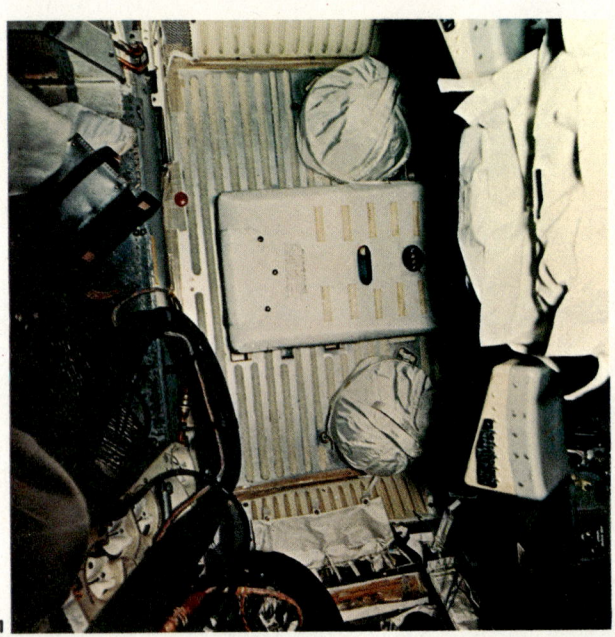
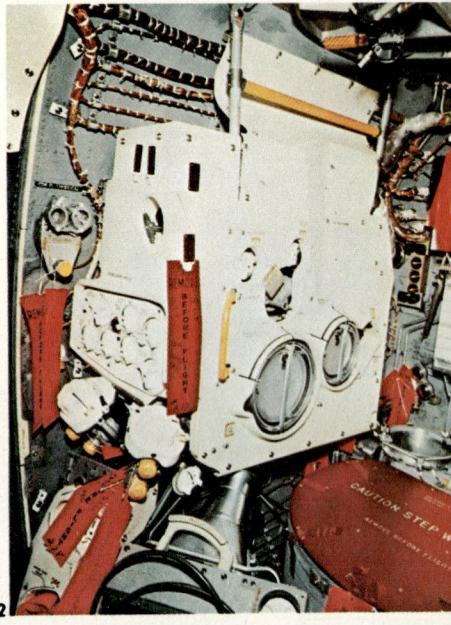

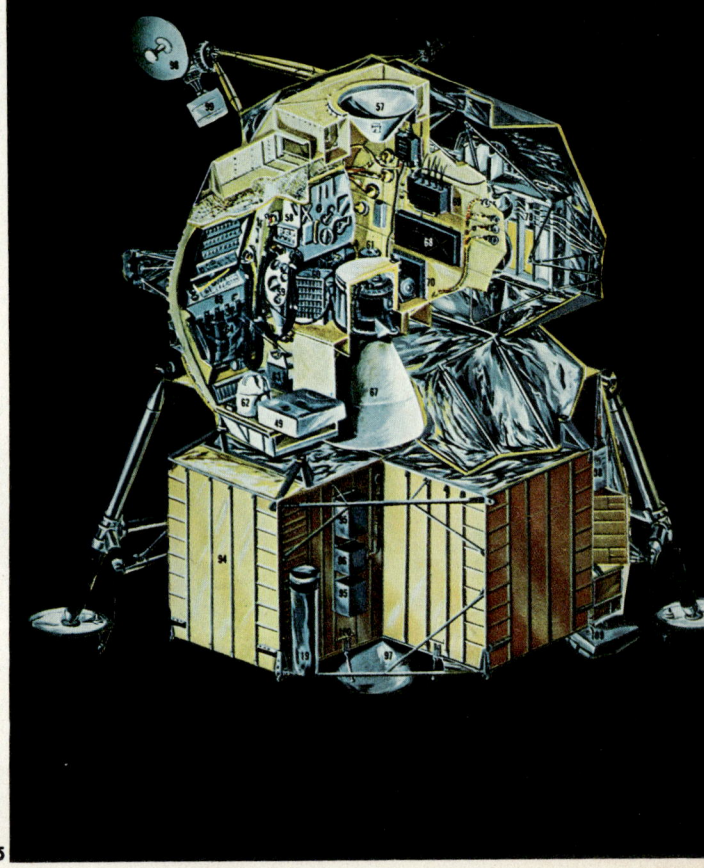

the Apollo 12 landing mission. Note the protective carpeting covering the floor of the cabin, which will be removed before launch. To the right is the ECS pack, while at left are stowage packs for food, liquid cooling garments, and inflight data cards. Bean looks down at the suit liquid cooling assembly unit. Attached by hoses to the pressure suits while inflight, it maintains comfortable body temperatures. A series of NASA/Grumman renderings (**4,5,6,7**) show the structural arrangement of the LM in cutaway form. In the first view (**4**) is seen the familiar front face, but stripped of its protective foil coverings. The landing legs are extended, but the long landing sensor probes are retracted. Directly in front of the forward hatch at left is the "porch." To leave the LM, each astronaut crawls out of the hatch onto this platform, then descends to the surface by the leg-mounted ladder. The second view (**5**) shows the outer skin removed. At top center is the funnel-shaped docking drogue, into which the probe on the CM fits. After docking, this mechanism is then removed, and the aperture becomes the tunnel, accessible by a hatch. The next view (**6**) is looking forward from aft, with the docking hatch swung open at top. The last view (**7**), looking upwards, is dominated by the descent engine's nozzle extension and the landing legs. These views are of an early LM, and do not show provision for the Lunar Roving Vehicle planned for the last Apollo missions. With the first spaceship, man has truly entered the Age of Space Travel.

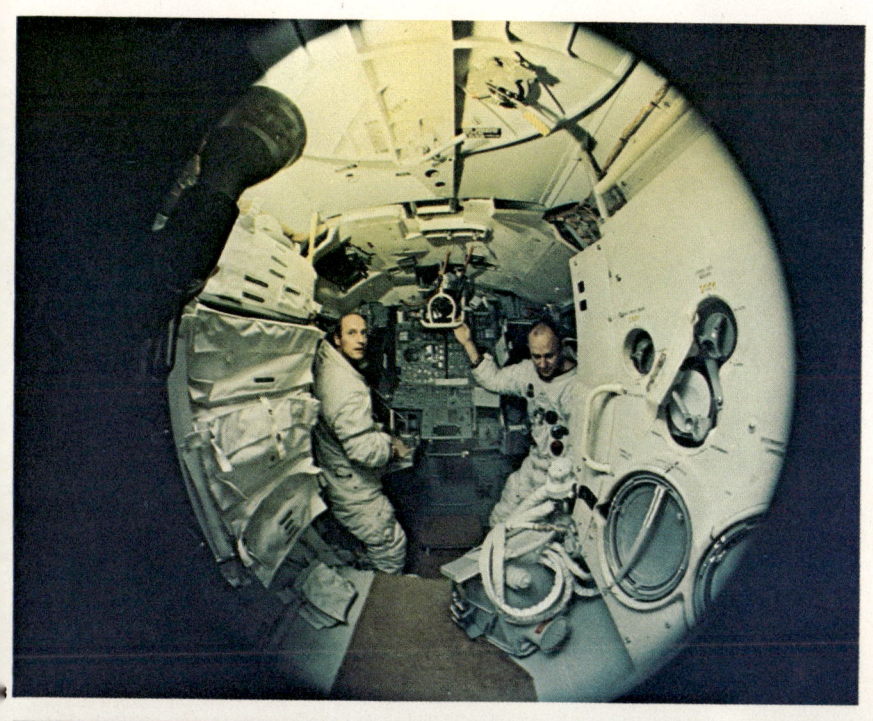

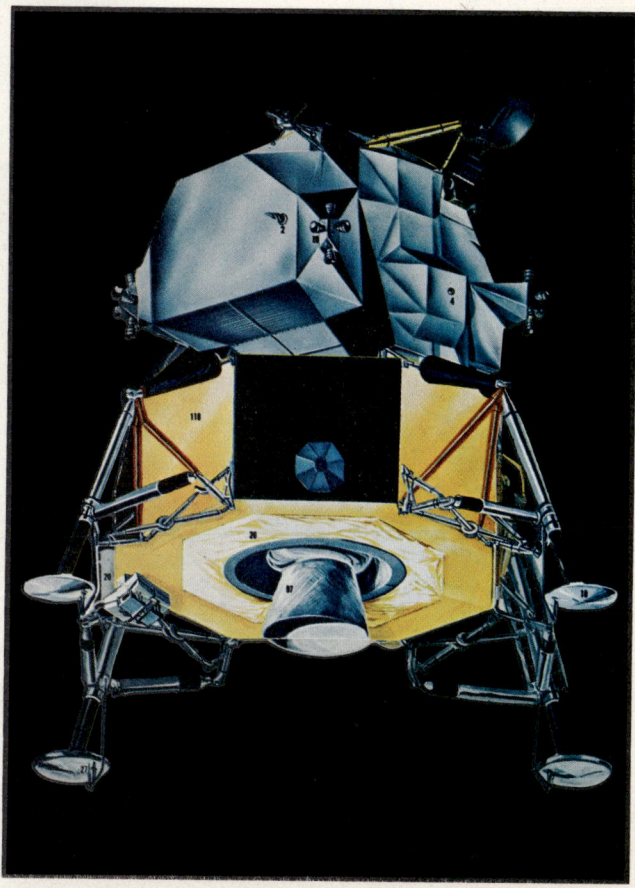

APOLLO 9

Several factors make Apollo 9 among the most significant steps in mankind's increasingly—bolder advance into the new frontier of space. Though this Apollo flight will not reach out into deep space, it is a milestone effort, a mighty dress rehearsal for the near future. For the first time, the Lunar Module, the vital fragile-looking vehicle that will land men on the moon, will be manned and flown. The crew of McDivitt, Scott, and Schweickart train as no crew has trained before, for they must perform maneuvers never previously attempted. They will disengage their Command/Service Module from the Saturn S-IVB stage, turn it around, link up with the LM, and pull it free from its compartment. The three modules that comprise the heart of the Apollo concept will then fly together for the first time. Next McDivitt and Schweickart will man the LM, detach it to fly alone, and practice lunar landing maneuvers. Then they must rendezvous and redock with the CSM, and transfer back to rejoin Scott. Schweickart is scheduled for a spacewalk, the first since Gemini 12. This crew is the first to name their spacecraft since the days of Project Mercury; in obvious references to their shapes, the CM is christened *Gumdrop,* and the LM becomes *Spider.*

JAMES A. McDIVITT

SPACECRAFT COMMANDER/ Born on June 10, 1929, in Chicago, Illinois. Colonel, USAF. Degrees: BS in Aeronautical Engineering, University of Michigan; Honorary PhD in Astronautical Science, University of Michigan. A former Air Force test pilot, Jim McDivitt has over 4500 hours flight time. He was command pilot on Gemini 4, which saw the memorable first American spacewalk. In June 1966 he became Manager for Lunar Landing Operations, responsible for planning all lunar flights beyond the first landing mission.

DAVID R. SCOTT

COMMAND MODULE PILOT/ Born June 6, 1932, San Antonio, Texas. Colonel, USAF. Degrees: Bachelor of Science from the U.S. Military Academy; BS and MS in Aeronautics and Astronautics from MIT. He attended both the Air Force Experimental Test Pilot School and the Aerospace Research Pilot School before being selected as an astronaut in 1963. He was the pilot on Gemini 8, which performed the first successful docking of two craft in space, and displayed great skill on that flight in overcoming a faulty thruster.

RUSSELL L. SCHWEICKART

LUNAR MODULE PILOT/ A former pilot for the USAF and Air National Guard, "Rusty" Schweickart was born on October 25, 1935, in Neptune, New Jersey. Another MIT alumnus, he earned there a BS in Aeronautical Engineering and an MS in Aeronautics and Astronautics. Then employed as a research scientist at MIT's Experimental Astronomy Laboratory, he resigned from the military when selected by NASA in 1963. The proud father of five children, he and his wife Clare enjoy many outdoor sports.

APOLLO 9
Pre-launch

At North American Rockwell's Space Division in Downey, California (**1**), the Apollo 9 crew prepare for a review of the hardware stowage arrangement on the spacecraft. At left is Russ Schweickart, center David Scott, and Jim McDivitt at right. All concerned with Apollo 9 train for the mission as never before. Every phase, every evolution and every plausible emergency is studied, tried out in simulators, and then re-evaluated. A short break comes at Kennedy Space Center (**2**), where barbers Matt Jude, foreground, and Chuck Quadrozzi give Scott and McDivitt a close cropping. Russ Schweickart, not shown, doesn't escape the shears, however! There is no time in space for any tonsorial luxuries. Back into the training routine (**3**) are David Scott, foreground, and Jim McDivitt at the Manned Spacecraft Operations Building in a simulated altitude chamber test. For every hour spent in deep space, the astronauts undergo far many more hours of extremely demanding training, practicing in full-scale mockups of the Lunar Module and

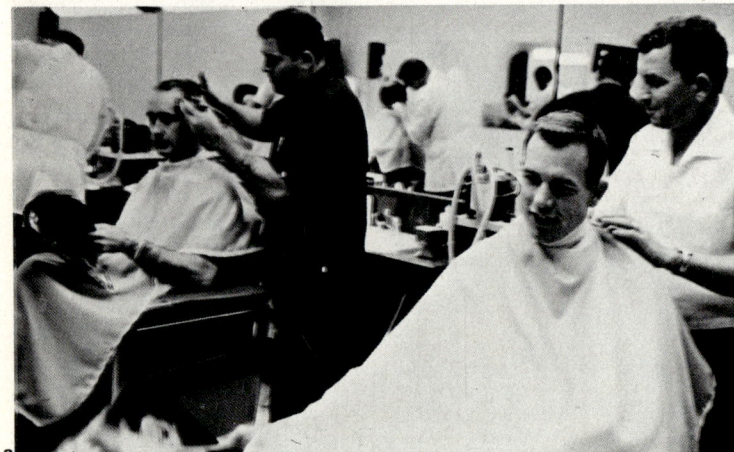

Command/Service Module. Training equipment has reached an incredible state of sophistication: the effect is carried to the point that an LM trainer, simulating a descent on the moon, actually casts a shadow on a simulated moonscape. The training vehicles, supported on columns of air to duplicate their motion in space, are linked to computers which are able to reproduce any phase of the mission. Seemingly heartless ground instructors create every possible form of emergency condition without notice, some bad enough to cause the immediate abort of a real flight. All three crewmen fall sick under the pressure—colds—but the real problem is utter mental and physical exhaustion. The men's bodies cry "Slow down!" After one too many 18-hour training days, Dr. Charles Berry clamps down and for the first time a launch is rescheduled three days later. A costly delay, but the Chief Flight Surgeon states that three sick astronauts in orbit could prove far more costly. Crew members have a history of post-flight "letdown," but Apollo 9 is too critical for any chances. Meanwhile, in High Bay 3 of the Vehicle Assembly Building, the 138-foot-high first stage of Apollo 9's Saturn V launch vehicle (A/S 504) is hoisted (**4**) in preparation for stacking on the Mobile Launcher. Except for some minor difficulties, confidence in the mighty launch vehicle is high at the Cape. Each stage is burning slightly overtime, and aggravating yaw and oscillation problems are evident, but none will halt Apollo 9. The launch is now scheduled for March 3.

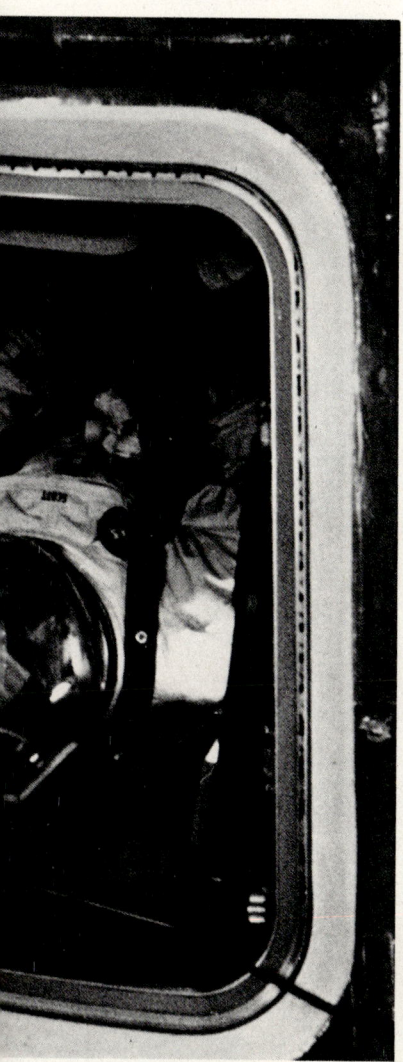

APOLLO 9
Launch

APOLLO 9
Mission

Eleven minutes after launch, Apollo 9 achieves its intended orbit, a nearly circular one at 103 nautical miles in altitude. The time is 11:11 AM, EST, March 3, 1969. Beneath them, the crew sees the panorama of the Florida peninsula (**1**), looking due south; clouds are scattered like cotton puffballs over the ocean. But there is little time to gaze at the earth; the crew has the toughest in-space schedule to date. Three hours after lift-off, *Gumdrop* is pulled clear of the third stage by explosive bolts. The Lunar Module, *Spider*, then has its protective shielding shell blown clear (**2**). Now, for the first time, a CSM is maneuvered around in a 180-degree tumble until its nose is pointing towards the Saturn third stage. For this maneuver, the CSM's small attitude control rocket clusters are fired, "pitching" the CSM around. *Spider*, its landing legs retracted, is seen through *Gumdrop's* docking window (**3**). Scott's target is the black cone at the center of *Spider*; this is the docking drogue to which *Gumdrop* will attach itself. It is through this

is spent finalizing the docking maneuver. Again, the countless hours of practice are going to pay off. But practice is over, this is the real thing. The flight simulators are hundreds of miles below them. With patient skill, Scott aligns the CSM with the LM, holding off at about 75 feet. Carefully, he aligns the docking target on the LM with the reticle crosshairs in his periscope-like alignment

It is for these maneuvering conditions that the 12 small attitude control liquid-fuel rocket engines are mounted in clusters of four around the hulls of the CM and SM. Each of these engines can be started a total of 10,000 times during a mission, and can fire for as short a period of time as 12 milliseconds. Thus, extremely delicate and precise maneuvers are within the range of the

firmly joined. Spring-loaded rams now push the LM out of its compartment, and the linked modules have now completed this critical maneuver which will be performed by all future Apollo missions. Once clear from the third stage, the crew gives Mission Control the word, and the S-IVB's J-2 engine is restarted by ground command, sending it far away to an eventual fiery end in the sun.

APOLLO 9 Mission

After the docking is completed, and the third stage has been disposed of, the linked vehicles begin a series of tests. The SM engine is fired both to extend the orbit life and to examine the effects of the firing on the docked vehicles. Scott informs Mission Control that they can feel vibrations throughout the craft, but there is no mishap. The digital autopilot is checked out by firing the SM engine at angles. Everything checks out OK. The fuel consumption lightens the CSM which would enable a rescue of the LM easier to be carried out later in the mission, if necessary. In such a situation, the small reaction thrusters would be sufficient to maneuver the CSM. On the third day, the LM crew enters *Spider*; the vehicle is pressurized and its power turned on. Now the punch is reversed: for the first time in space, the throttling Descent Propulsion Stage engine in the lower half of the LM is fired, both under manual and automated control. This vital piece of power will enable astronauts to land safely on the moon's sur-

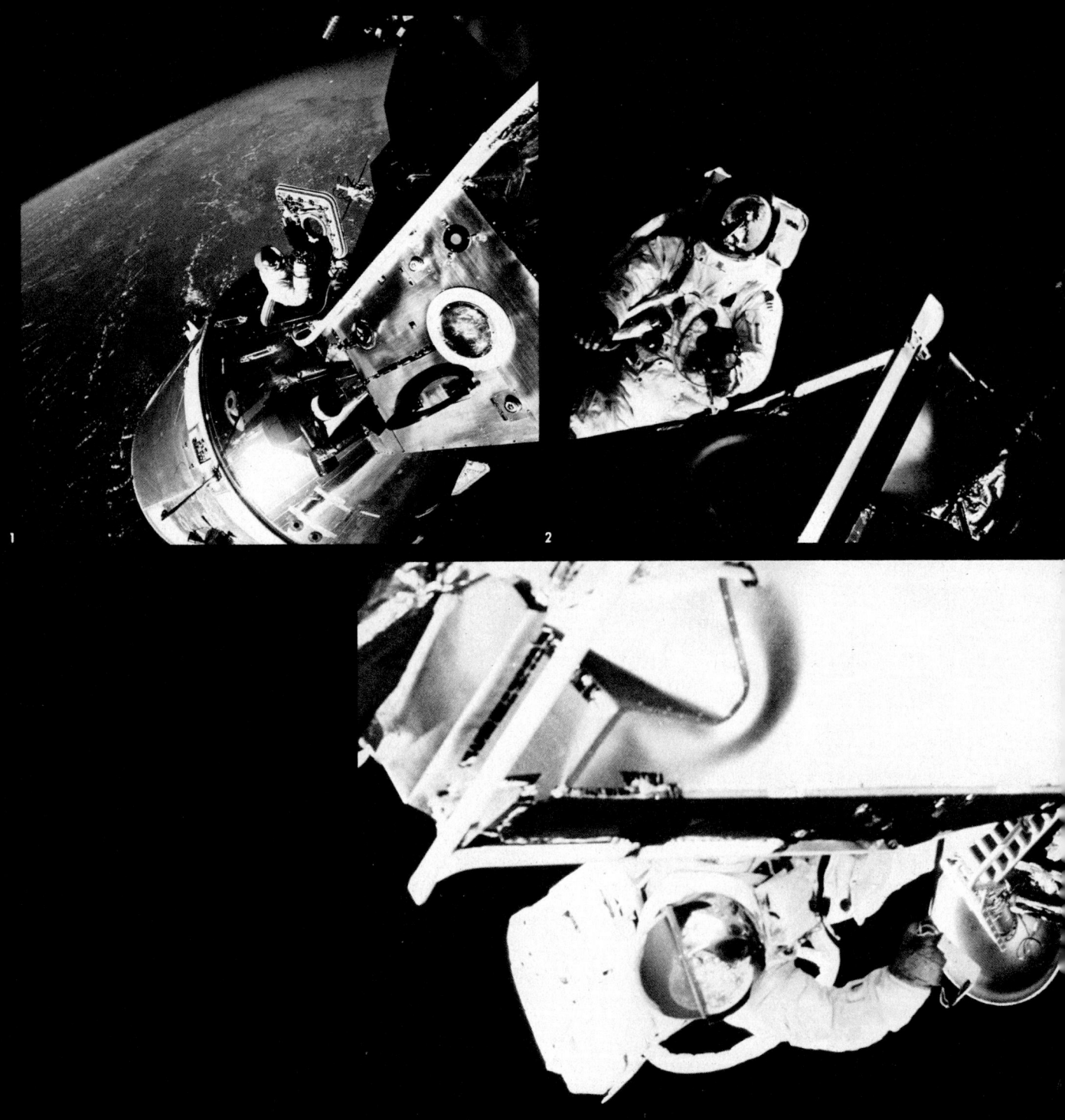

face. Again, all systems remain go. Now McDivitt and Schweickart return to the Command Module via the docking tunnel. Schweickart has been experiencing bouts of nausea, draining his strength; all three are still feeling the effects of their last-minute training program. Momentary fears are overcome as Schweickart quickly recovers, but Scott moves the planned EVA up by one day, and cuts the period to 38 minutes. A burn from the CSM engine fixes the two vehicles' orbital height at 135 miles. On the third day, the EVA begins. McDivitt and Schweickart re-enter the LM. Then both spacecraft are depressurized and their hatches opened. Schweickart, who is laughingly code named "Red Rover" for the EVA, photographs Scott (**1**) in the CM hatchway. "Boy, oh boy, what a view," exclaims Red Rover, over the wide-open panorama of earth. The bright handrail is clearly visible, though Red Rover does not use it fully, as a hand-over-hand walk has been curtailed. Scott returns the favor, and shoots Schweickart who is standing in the LM hatchway (**2**) using the foot restraints—the "golden slippers"—to maintain his position. Reflected in his visor is the CSM, and below, the cloudy face of earth. Schweickart moves around the side of the LM (**3**). He is wearing the Portable Life Support System, a duplicate of that to be used on the moon. Floating a few feet from the LM (**4**) Red Rover catches this view of the two craft, while close in, he photographs Scott (**5**). "Gotcha!" says Schweickart (**6**) on *Spider*'s porch.

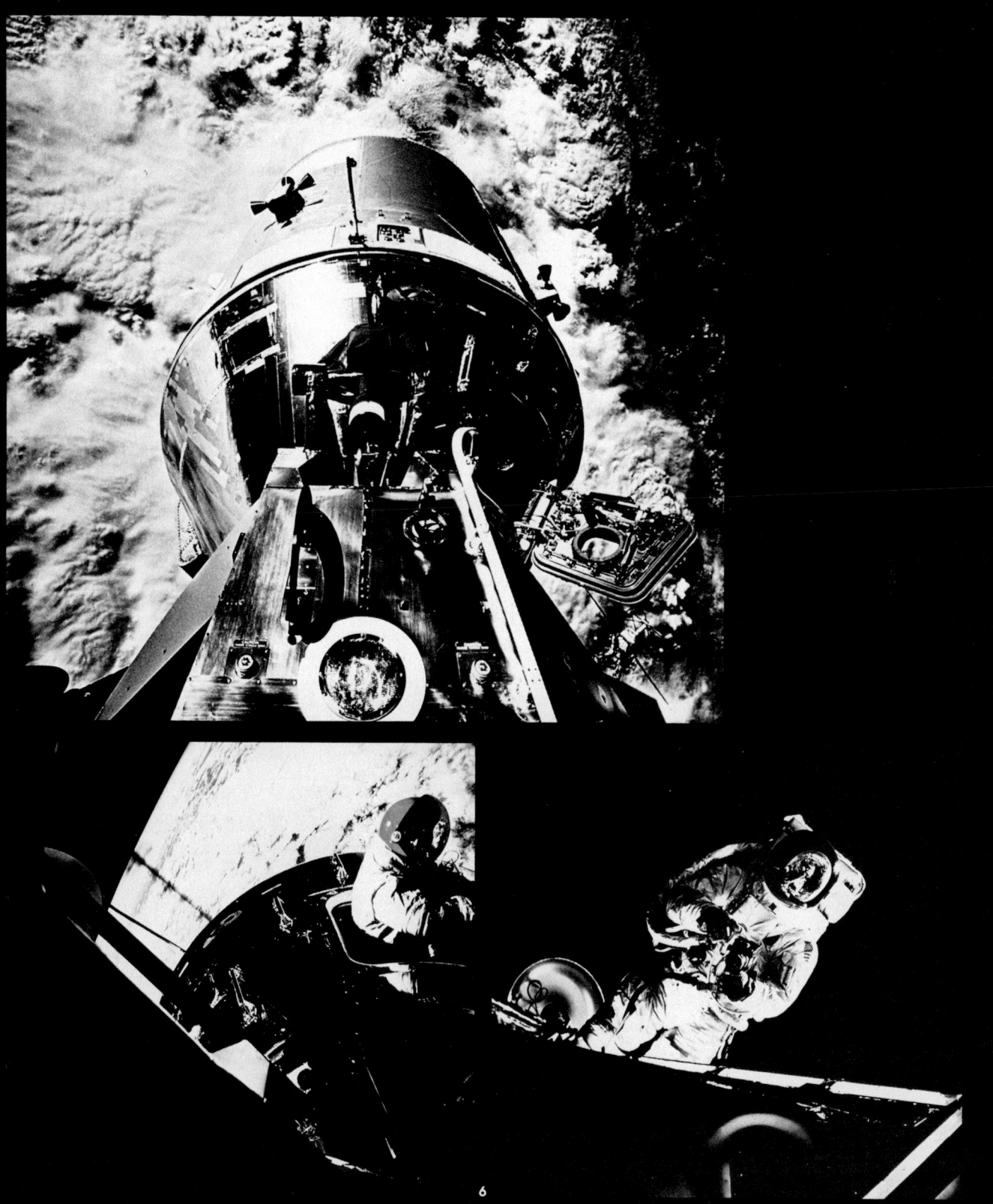

APOLLO 9 Mission

During his EVA, Schweickart seems to forget his problems of the previous day and enjoys the sensations of free fall. While using the handrails on the LM (**1**), he comments that his hands seem slightly warmer than the rest of his body, but not uncomfortably so. His only link with the LM is a tether line. Finally, Scott calls in Red Rover, who was enjoying his new environment more and more. It has been the first space walk for our astronauts in two years. But there is more work to be done, so McDivitt and Schweickart transfer to *Spider*. The cab is pressurized and power turned on. Cautiously the two vehicles undock, and for the first time ever, an LM flies alone in space. *Spider*, looking very much like its namesake, executes a superb roll in space for Scott (**3**). "You are upside down," he comments laconically. "I was just thinking, one of us isn't right side up," retorts *Spider*. In space, vertical and horizontal have little meaning! The LM crew blip the descent stage engine, but only enough to move *Spider* three miles away from the CM.

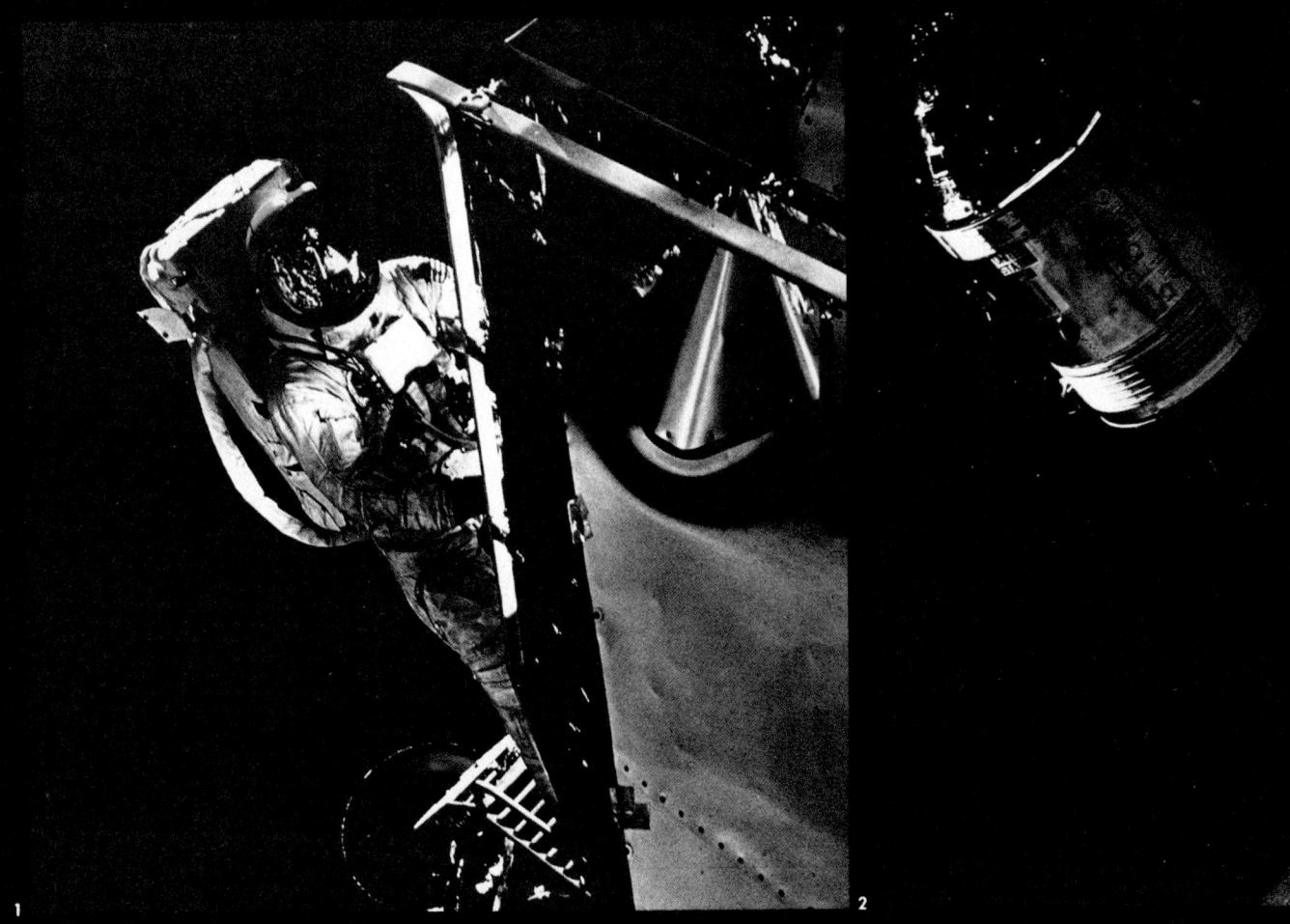

Two further orbits are spent with the two spacecraft this close together; in case of trouble with the yet-unproven LM, Scott will be able to move in quickly to the rescue. A second blast from the descent engine shoves the LM 11 miles from *Gumdrop* for further instrument and flight testing. The third burn of the LM engine drops it 100 miles behind the CSM and out of visual contact. Unfortunately, its main tracking strobe light has gone out, but *Spider* is able to see the CSM as a bright spot as far as 49 miles away. Close in, *Gumdrop* is seen from the LM (**2**), and shows its operational scars from the hull-mounted reaction control engines. Scott maintains radio contact with *Spider*, but remains passive. *Spider* undergoes a full set of maneuvers similar to those to be made during an actual descent to the lunar surface. In a descent attitude (**4**) *Spider* drops a short distance under the eye of *Gumdrop*. Its delicate frame could never allow it to operate close to the powerful grip of earth's gravity, much less in its dense atmosphere. Finally, the descent stage nears the end of its life. Its fuel load is almost expended in the evolution of the simulated descent and maneuvering operations. The eight-hour test session high in the sky has proven that the LM can carry out the mission it was designed to do: land men on the moon. Only one tricky job is left to do: jettison the descent stage and return to the CSM with the ascent engine, using radar to accurately rendezvous and close in. Then *Gumdrop* will perform the final link-up.

APOLLO 9 Mission

Spider, seemingly black, hangs in space (**1**) with its spindly landing legs extended; legs that will never touch the lunar surface, but have helped pave the way to the moon. Now explosive bolts are armed in order to separate its descent stage from the ascent stage. When fired, they drive guillotine-like blades through every pipe and connection between the two stages. As Schweickart fires the ascent stage engine, the descent stage is blown away. Its fate is a searing end in earth's atmosphere, unlike future descent stages that will remain in silent repose on the moon forever. Using their on-board radar, the LM crew quickly find their way back to *Gumdrop*. As they approach, lining up by sight (**2**), the intense flare of the sun makes alignment difficult. Scott tells McDivitt and Schweikart that they're "The biggest, funniest-looking spider I've ever seen!" *Spider* is different looking with its lower half gone. Again *Spider* goes through a somersault for *Gumdrop* (**3**), showing its bottom and sides for Scott's inspection. The protec-

tive Mylar shielding shows some bending from the blast of the ascent engine, but all remains intact. Above the engine can be seen the "porch" from the hatchway (**4**). The CM's reaction rocket clusters are used again to align the two craft for docking, simulating a future event 69 miles above the lunar surface when returning veterans from the lunar surface will reconnect with their CSM. Tension reaches a high point as the two craft make the final closing approach from 100 feet apart. "One of us isn't right side up," says *Spider*. "I can't see you in there, Dave." "Oh, I'm in here," replies *Gumdrop*. *Spider*: "Let's stick together." *Gumdrop*: "I'm with you!" Once again, both crews use their COAS optics, carefully aligning the docking targets as they come together. Before they know it, the contacts close and the docking signal sounds. *Spider* and *Gumdrop* are back together, and Apollo 9 has just about earned its laurels. "Okay, Houston, we're docked!" informs Scott. The above dialogue, carried out during one of the most important phases, belies the tenseness of the operation. Schweickart asks for three days bed rest after the flight, and the Capcom in Houston immediately replies, "Three days off." After the LM crew is back aboard the CM, the *Spider* is undocked, and a radioed signal from the ground fires its engine, sending it hurtling off into eternal retirement in deep space. Mission Control asked if they forgot the LM Pilot. "I didn't forget him," chuckles McDivitt. "I left him inside *Spider* on purpose!" It is March 9, 1969.

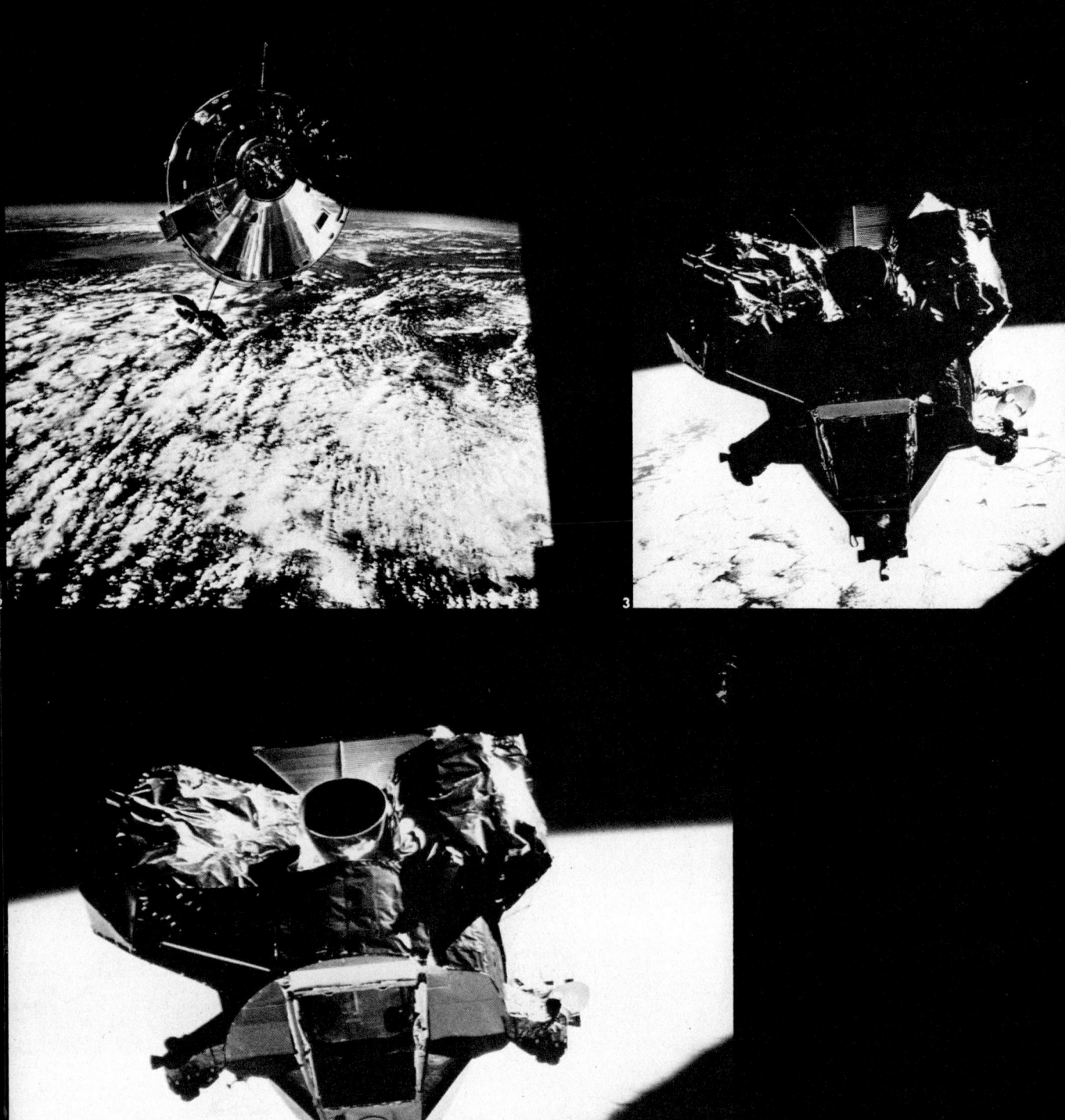

APOLLO 9
Splashdown

The remaining five days of the mission are filled primarily with photographing earth surface features and tracking. We have only begun to realize how valuable this super-high-altitude picture-taking can be. Earlier space photographs revealed geologic formations and panoramas that scientists were not aware of. Orbiting spacecraft have given us a new perspective of our earth. *Gumdrop* is equipped with four 70mm cameras capable of working with not only black-and-white film, but infrared and color infrared. Infrared pictures, among other uses, show the condition of timber and crop resources and penetrate the ocean's surface, suggesting the presence of oil and mineral deposits beneath its shallows. At last, the time for re-entry and splashdown arrives. A nasty storm in the preselected splashdown region results in a shift of sites 500 miles to the south. But new navigation and tracking techniques allow the USS *Guadalcanal* to be a mere three miles from the Command Module as it strikes the ocean. Resting on its flotation

collar, *Gumdrop* rides easily as pararescuemen escort the three crewmen to a raft (**1**). Televised closeup for the first time, a scene of unwanted slapstick ensues—the downdraft from the rescue helicopter's rotor overturns a raft. Result: McDivitt and Scott receive their first bath since lift-off—in cold and sticky saltwater while in the recovery basket hanging from the guilty chopper. A warm welcome, however, awaits them on the flight deck (**2**). Schweickart and Scott look little the worse from a distance, but Jim McDivitt has sprouted quite a growth in 10 days! TV cameramen crowd their way down the red carpet. The big 'copter (**3**) pulls all that remains of *Gumdrop* from the water, while its parachutes lie like huge wet handkerchiefs in the water. The CM will eventually go back to its birthplace at the North American Rockwell plant in Downey, California for a full post-flight checkup, but will not be reused. After physical checkups and debriefing, the Apollo 9 crew (**4**) Jim McDivitt, Dave Scott, and Russell "Rusty" Schweickart, left to right, hold a press conference. The feeling from all sides is one of pride as well as knowledge of success. Apollo 9 has opened the way for another *Spider* to go into space, this time all the way around the moon in the final step before an actual lunar landing. The crew receives greetings from the next generation of astronauts (**5, 6**), both boys and girls. The "next time" in space will be too soon for them, but the future holds many bright promises. The moon is much closer now, within our reach.

Before man learned to fly, the face of his earth fascinated him. The trouble was that he couldn't look down upon it. A view from the tallest mountain summit was always restricted by the continuing peaks. Man ached for the freedom of a bird. In 1817, a German, Puckler-Muskau, ascended over Berlin in a balloon and was ecstatic with the view. Although ballooning became a passion to the few who could afford it, the first aerial photograph didn't appear until a Frenchman, Nadar, breathlessly pulled it off in 1858. Those were the days of the wet-plate process of developing, which required completion within 20 minutes after exposure—a not-too exciting prospect for a balloonist. A more sophisticated balloon camera was patented by the Englishman Walter Bentley Woodbury in 1877, which was able to take no less than five hundred 9 x 9-inch pictures on a single roll of film; it was operated by a cable from the ground. At about the same time, cameras were being harnessed to kites, and even to the chests of carrier

SPACE PHOTOGRAPHY
A new breed of cameras for a new kind of tourist

PHOTOS COURTESY HASSELBLAD

pigeons! Needless to say, lost cameras were not uncommon. But in 1912, the German Army had in use a 20-foot-long photo rocket, powered by gunpowder. The rocket was fired to an altitude of about 2600 feet in eight seconds and then the rocket's nose cone, containing a camera, was ejected. Taking pictures by timer, the camera dropped by parachute. Airplanes by that time had reached altitudes of 12,000 feet, and aerial photography became important in World War I military operations. Photo reconnaissance by the Allies in World War II greatly assisted strategic bombing plans. Military sophistication of Photo-Recon achieved such maturity during the war that postwar scientific and engineering applications mushroomed into a firm foundation for space camera wizardry. The celebrated Lockheed U-2 spy plane was able to take pictures regularly from 90,000 feet in 1960, but television pictures had then appeared from America's Explorer VI satellite at an altitude of 429 miles. When the time came for an American to enter space, John Glenn's hand-held Minolta High-Matic 35mm camera was thought suitable (**1**), due to distance and cargo limitations. But in October 1962, Hasselblad, of Goteborg, Sweden, was designated the official NASA camera manufacturer (**2**). An assortment of long lenses, of course (**3**), is highly desirable. At present, the motor drive 500EL/70 model is the standard space camera (**4**), and another variation is the "Lunar Surface" 500EL/70, with its Zeiss lens (**5**). A smaller version is the 500C (**6**).

PHOTOGRAPHY
Challenge

With the appearance of satellites and later of astronauts, altitudes which were once discussed in *feet* now became *miles*. The following series of photos shows a progression from a 10,000-foot aerial view of the tiny fishing village of Casma, Peru (**1**), to a high-altitude infrared shot of Washington, D.C. (**2**), taken from 50,500 feet. Also note an Air Force RB-57 view of Raritan Bay, New Jersey (**3**). The photo was taken at 61,000 feet by an Air Force RB-57 reconnaissance plane, using an RC-8 camera with Aerial Ektachrome film and Haze Filter #3, plus 2.2 anti-vignetting. Quite a contrast, these, to Lunar Orbiter II's bleak image, taken from 27.8 miles above the lunar surface (**4**). A beginning of studies on lunar landing sites, the photo is a single picture composed of 27 of the 86 framelets which make up a complete telephoto frame. At this point, man's ability to take pictures above his world had progressed from a balloon to the moon, in just the last 108 years of his million-year existence. Aerial photography

had come a long way, from Nadar's 1858 ascent of 240 feet above Paris, to Lunar Orbiter II's 1966 moon surveillance at 240,000 miles above his planet. The Lunar Orbiters contained both a photographic and a television system. There were five such satellites in all, each having two lenses: a 24-inch telephoto and a 3-inch wide-angle lens. Each had apertures of f/5.6 and shutter speeds of 1/25th, 1/50th and 1/100th of a second. The danger of image movement and thus of blurring was negated by a sensing device—developed in aerial photography—which compensated for image movement of the film during exposure, in the same direction and at the same rate as the image. Processing, however, was even more intriguing. The 70mm special high-definition aerial film was contained in one continuous loop around 30 individual rollers; contained within the spacecraft was complete processing, printing, drying, and readout-transmission capability. A finished photo was on earth in less than 15 minutes following original exposure (and was eventually reconstructed on 35mm film). Still, the challenge to land on the moon dazzled man's imagination, and led to more cameras. The Cloud Top Spectrometer (**5**) was used on Gemini 5 to measure various cloud formations. Another development, the Hasselblad 500C (**6**), was designed to take a photograph every six seconds of a 100-mile-wide area of the lunar surface. A cluster of Hasselblad 500ELs (**7,8**), give a multi-filter, multispectral look at a particular subject.

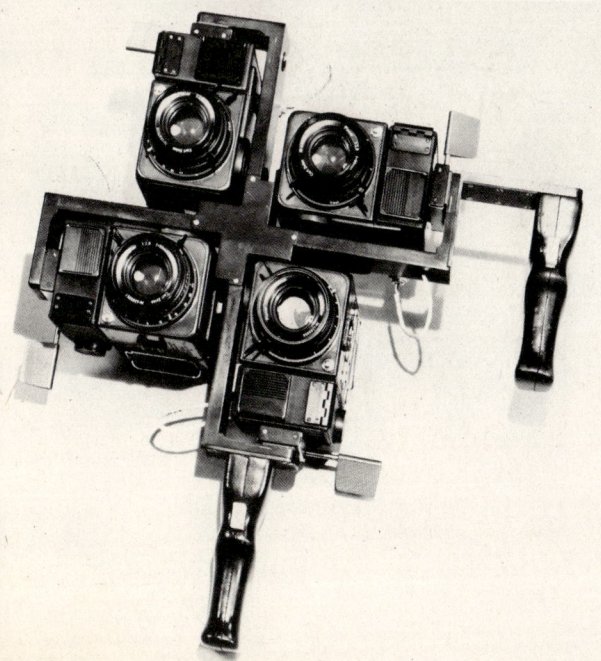

PHOTOGRAPHY
Early missions

The early manned space missions were concerned with obtaining and testing the best cameras for future missions. NASA's high standards and quality demands resulted in several modifications—even to the smallest Hasselblad. To begin with, NASA didn't allow glass of any kind on a spacecraft, due to the possibility of breakage from changes in pressure and temperature (although special exceptions were allowed for windows and lenses). This meant the removal of the Hasselblad's mirror and reflex viewfinder. Leatherette camera coverings were removed, as was chrome trim which reflects glare in the exceedingly bright light conditions of space. There was also a change of camera lubricants, since conventional lubricants might easily ignite in the pure oxygen atmosphere of a pressurized capsule. An attentive listener to this situation is Mercury-Atlas 8 astronaut Wally Schirra, during a briefing by NASA's "Red" Williams, as Deke Slayton looks on (**1**). Schirra was the first astronaut to use a Hasselblad in space. The

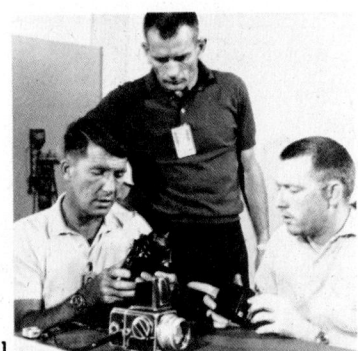

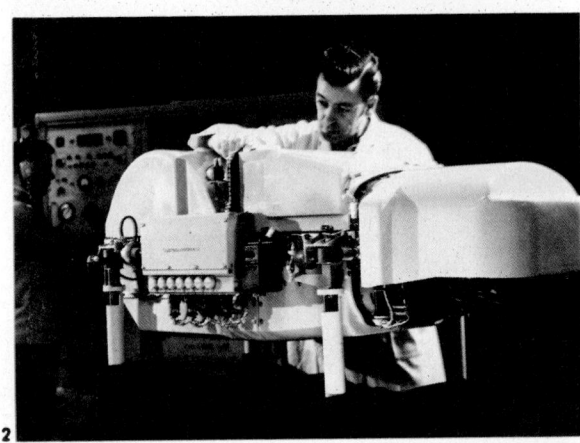

IMPERIAL VALLEY PHOTOGRAPHIC EXPERIMENT FOR CROP IDENTIFICATION

STUDY OBJECTIVES
EVALUATE
- SYNOPTIC SCALE PHOTOGRAPHY
- MONTHLY CROP CHANGES
- MULTISPECTRAL PHOTOGRAPHIC WAVELENGTHS

IMPERIAL VALLEY CHOSEN FOR STUDY BECAUSE
- AVAILABILITY OF GEMINI AND APOLLO SPACE PHOTOS
- ALL YEAR GROWING SEASON
- TOTALLY IRRIGATED AGRICULTURE
- VARIETY OF IMPORTANT CROPS
- TOTAL AREA AND FIELD SIZE LARGE ENOUGH FOR SYNOPTIC EVALUATION

GENERAL INFORMATION
- 450,000 ACRES
- TOTALLY IRRIGATED
  - 2.5 INCHES RAINFALL/YEAR
- MAJOR CROPS
  - BARLEY
  - SUGAR BEETS
  - COTTON
  - ALFALFA
  - SORGHUM
  - VEGETABLES

GEMINI V PHOTO
AUGUST 21, 1965

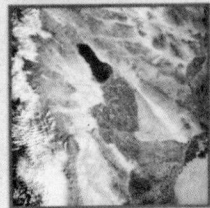
APOLLO 9
COLOR PHOTO
MARCH 12, 1969

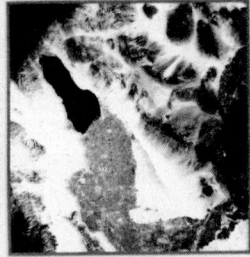
APOLLO 9
COLOR IR PHOTO
MARCH 12, 1969

biggest of the space cameras, though, is the Itek Optical Bar Panoramic camera, (**2**), which contains, in a massive film cassette (under the technician's right elbow), a film roll which is five inches wide and one mile long! It will take 1500 individual pictures. At 60 miles altitude above the moon, this KA-80A allows identification of lunar objects as small as 54 inches across! Its f/3.5 lens is attached to a gimbal structure which constantly revolves, giving motion compensation to its two-dimensional and stereo capability.

By the time of Gemini 11's 26th orbit, pictures such as this Middle East overview (**3**) were commonplace. At an altitude of some 500 miles, the Red Sea, the Sea of Galilee, and the Mediterranean can be seen clearly. To understand proportions better, the photo scale effects of space photography can be measured (**4**) in this composite of California's Imperial Valley. Using infrared film which can detect the heat of growing plants, space cameras provide a stunning example of monthly crop change measurement (**5**).

In this black and white photo, infrared film shows crop growing areas as a dark, meandering ribbon, connected to the Salton Sea; in color processing these areas would be a bright red. Here is another view of the Hasselblad 500C (**6**). Note the cross-hairs for sighting. Another camera, this one to track and photograph terrestrial objects, is the 35mm Zeiss Contarex single lens reflex (**7**). The elaborate tracking lens and timer are shown. There will also be a lunar stereo close-up-flash camera for use on the moon.

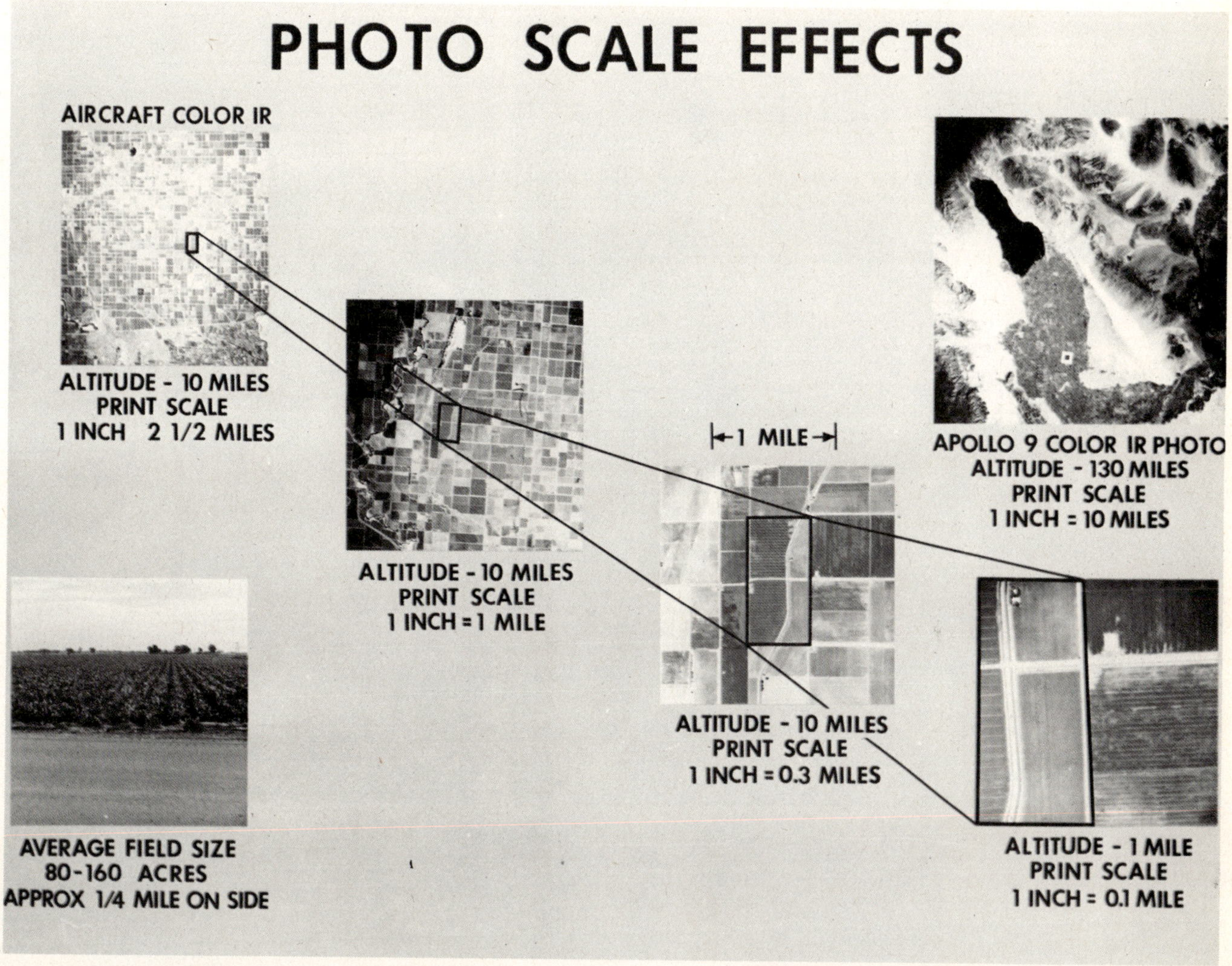

PHOTO SCALE EFFECTS

AIRCRAFT COLOR IR
ALTITUDE - 10 MILES
PRINT SCALE
1 INCH 2 1/2 MILES

ALTITUDE - 10 MILES
PRINT SCALE
1 INCH = 1 MILE

|← 1 MILE →|

APOLLO 9 COLOR IR PHOTO
ALTITUDE - 130 MILES
PRINT SCALE
1 INCH = 10 MILES

ALTITUDE - 10 MILES
PRINT SCALE
1 INCH = 0.3 MILES

AVERAGE FIELD SIZE
80-160 ACRES
APPROX 1/4 MILE ON SIDE

ALTITUDE - 1 MILE
PRINT SCALE
1 INCH = 0.1 MILE

4

6

7

PHOTOGRAPHY
Later missions

If you could make believe that you were an invisible passenger on Apollo 9's 151 orbits, you would be on hand for the true drama of space photography. On this mission are two Hasselblad 500C's, a Hasselblad SWC, and four Hasselblad 500EL/70's. During David Scott's fourth-day EVA, Lunar Module pilot Rusty Schweickart uses one of the 90-degree view angle Super Wide 500C's to magnificent advantage, as Scott emerges from the hatch of the Command Module *Gumdrop* (**1**). Scott then returns the favor, catching not only Schweickart on the porch of *Spider* (**2**), but in his visor, a somewhat distorted picture of the Command Module and the cloud-covered earth. The Hasselblad 500C uses an 80mm f/2.8 Zeiss Planar lens; the updated Super Wide C (SWC) uses a 36mm f/4.5 Zeiss Biogon lens. The cameras were able to capture the moon's forbidding-looking surface dramatically during the Apollo 8 Christmas mission (**3**). One of the candidates for the first lunar landing is the crater Goclenius in the Sea of Fertility

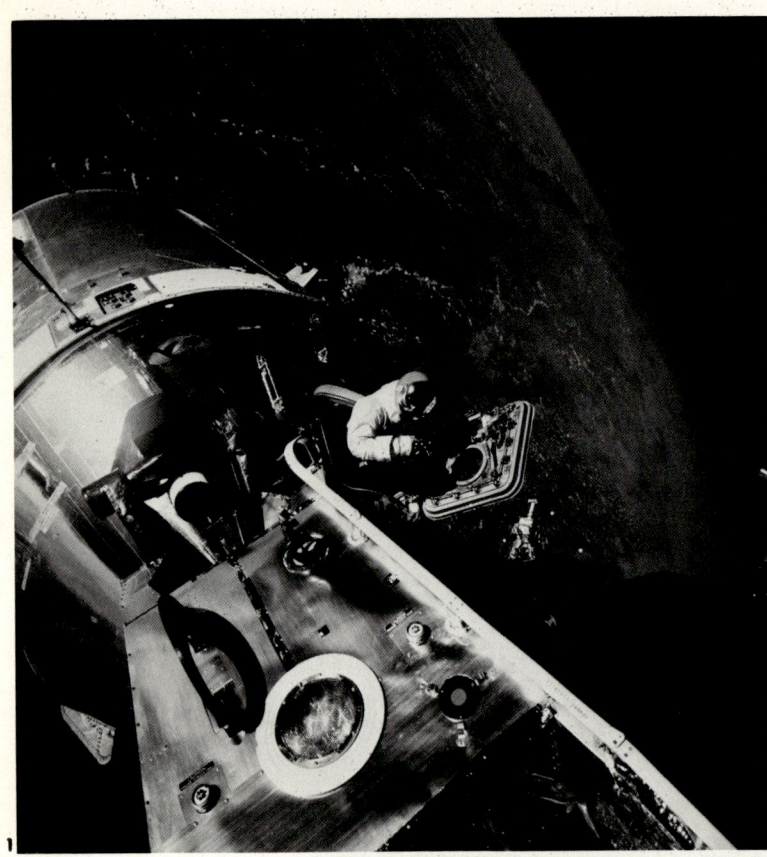

(**4**); and for this crucial selection process, a 250mm telephoto lens was fitted to a pair of 500EL's for maximum surface detail. As technical as this data may seem, the astronauts worry about it least of anybody. They are given complete instructions on various cards as to exactly what they're to photograph and what lens openings and shutter speeds to use. By Apollo 11 space photography will be into high gear. Later, with Apollo 15, a Hycon Mapping Camera will be installed in the SIM Bay of the Service Module to supplement high-altitude coverage. Both this and the Itek 24-inch Panoramic camera will require an EVA on the return trip to earth to retrieve their exposed film cassettes. For personal use (time permitting), a Nikon F 35mm camera is fitted with a 55mm lens and provided with four film cassettes. The strength of the thinner polyester-base film allows 200 exposures per black and white cassette and 160 for color. While the crew of Apollo 12, Pete Conrad (left) and Alan Bean, are busy in Arizona simulating a lunar walk (**5**), it is discovered that the Russians are having problems matching NASA's precise television resolution of the moon's surface (a revelation from the unmanned lunar probe Zond 6). Meanwhile, Conrad carries a gnomen, a sighting device for true vertical aiming. Both men carry the Fairchild 16mm Maurer movie camera; and later (**6**) at Cape Kennedy, suit-up to photograph simulated moon rocks. They have cameras with the Reseau plate (**7**) to aid in photogrammetric interpretation.

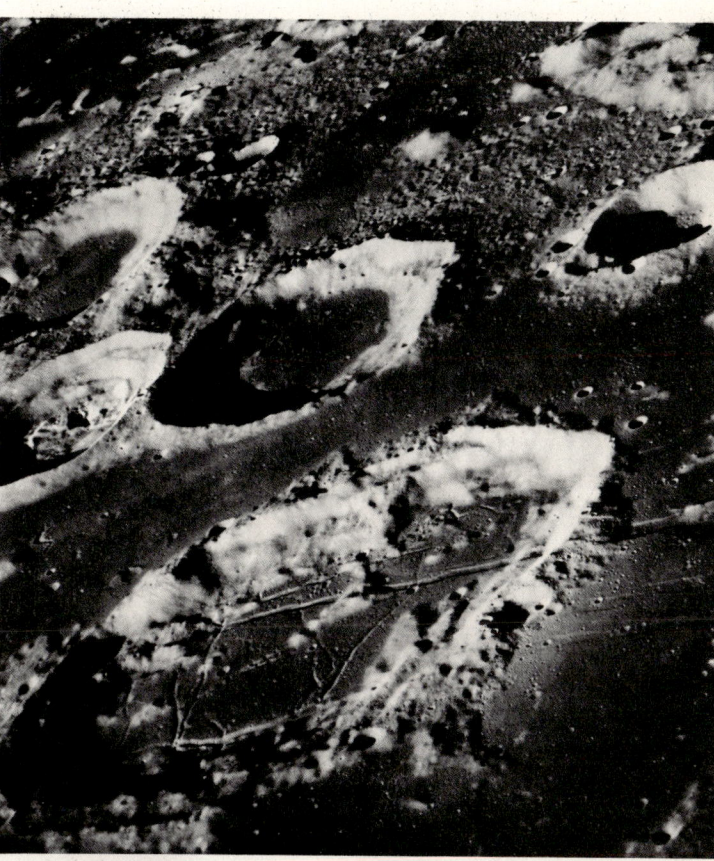

PHOTOGRAPHY
Future missions

Prior to Apollo 8, Jim Lovell was especially interested in the 16mm Maurer movie camera (**1**); its design is a mixture of esthetics and superb quality (**2**). For all Apollo missions, the camera types will be essentially the same, although on later missions, there will be more film. Planning ahead to Apollo 17, there will be nine cameras, including two Hasselblad 500EL's with Reseau plates; and 14 lenses. There will be eighteen 70mm film magazines; six in the Command Module, and 12 in the Lunar Module for use on the lunar surface. Apollo 17 will be a photographic feast, with enough standard NASA polyester-base film for 3600 color and black-and-white prints. Since the beginning, the astronauts have used extensively the fine-grained and high-resolution Panatomic-X Aerial Film (ASA 80). Also used is Kodax 2485 High Speed Recording Film—the most sensitive black and white available—with ASA flexibility of between 6000 and 10,000. Future missions will carry greatly improved color television cameras, of which

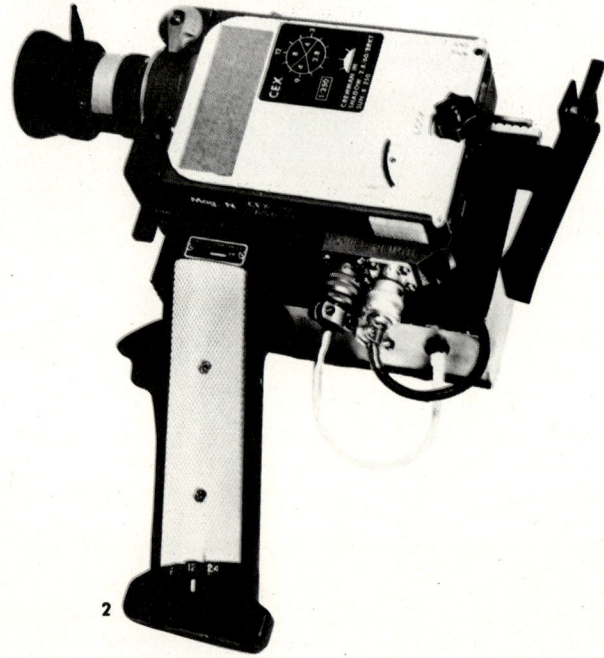

one is being tested here (**3**) in the Westinghouse thermal vacuum chamber. Once, on an Apollo mission, the wrong film was used; with the wrong exposure settings, the pictures should have been ruined. But NASA photo technicians, carefully shielding the returned film after word from the astronauts, began experimenting with identical film under simulated space conditions, until the right processing combinations compensated for the error. The real film was thus developed and its irreplaceable pictures saved. In the event of radiation damage to film, NASA knows precisely just how much to reduce development time in processing. The proof is always in the results, as demonstrates this successful low angle shot of the Fra Mauro area (**4**). Looking back on man's history with space photography, there has been gained a solid knowledge which will carry the dawn of his camera assault far into the universe. This Far Ultraviolet Camera/Spectroscope (**5**) is being designed to detect deep-space sources of hydrogen. Here, one of its lunar operators, astronaut John Young, operates a test at the Kennedy Space Center. Almost too soon, it seems, Young and Charles Duke will walk upon what seems to be, in this SIM Bay look, a truly desolate vista: Apollo 16's landing point in the Descartes region (**6**). For all missions, it's a new world of film; reassuring at launch with a television tracking camera (**7**). But space photography is greater than the sum of its parts. What is thus seen from space is held forever, for wonder and contemplation.

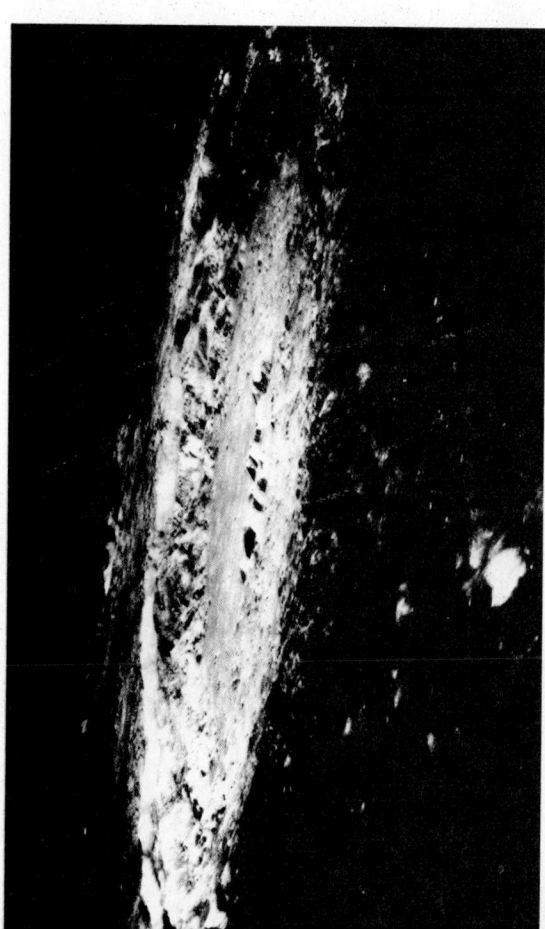

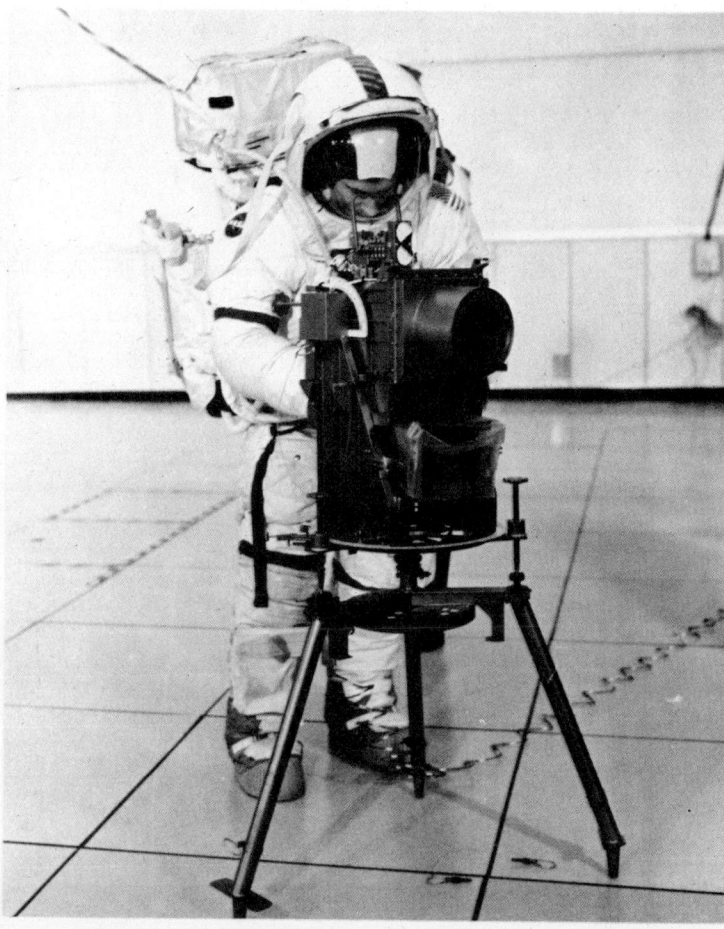

APOLLO 10

In January of 1969, the Soviets demonstrated that a crew transfer between two spacecraft in earth orbit was possible. Soyuz 4 and Soyuz 5 had superbly mastered rendezvous and docking, but the Russians decided that they were more interested in building a space station than landing on the moon. With the political pressure off, the United States could concentrate on perfecting a manned lunar landing. Apollo 9 successfully completed tests of moon landing techniques in earth orbit; Apollo 10 will go to the moon, and perform the same techniques from lunar orbit. But its Lunar Module will not land—this is the dress rehearsal. The flight has been set for May 18, 1969, with Thomas P. Stafford as commander, John W. Young as Command Module pilot, and Eugene A. Cernan as Lunar Module pilot. Stafford and Cernan are to descend to within 10 miles of the lunar surface, while Young remains in lunar orbit. The astronauts have named their spacecraft after the famous and much-loved characters from Charles Shultz's comic strip "Peanuts". Because the Lunar Module is supposed to "snoop around" scouting possible landing sites, the crew has named it *Snoopy*, and naturally the Command/Service Module then becomes *Charlie Brown*.

THOMAS P. STAFFORD

SPACECRAFT COMMANDER/ An Air Force Colonel, Stafford was commissioned in the USAF upon graduation from the Naval Academy at Annapolis. He was in succession a fighter pilot, test pilot, and instructor, and is the co-author of two widely used test pilot manuals; he has more than 6000 hours flying time. A NASA workhorse, he was backup pilot for Gemini 3, pilot for the Gemini 6 rendezvous mission, command pilot of Gemini 9, and backed up his Gemini 6 partner, Wally Schirra, as commander of the Apollo 7 flight.

JOHN W. YOUNG

COMMAND MODULE PILOT/ A "ramblin' wreck" from Georgia Tech, Young entered the Navy upon graduation and is now a Captain. He followed the familiar astronaut pattern of fighter pilot and test pilot before being picked by NASA in 1962. The pilot on Gemini 3, he then backed up Stafford on Gemini 6. He was command pilot for Gemini 10, a complicated flight which included rendezvous with two different vehicles, two EVAs, and a powered flight with a docked Agena. He has since been backup CM pilot for Apollo 7.

EUGENE A. CERNAN

LUNAR MODULE PILOT/ A Chicago product with a BS in Electrical Engineering from Purdue, Cernan entered the Navy through the ROTC Program. After a stint as fighter pilot, he earned an MS in Aeronautical Engineering from the Naval Postgraduate School. Picked by NASA in 1963, he has logged more than 4000 hours flying time. Since Gemini 9, where he and Stafford effected rendezvous with the "angry alligator" Agena target vehicle, he has served as the backup pilot for Gemini 12, and as the backup LM pilot for Apollo 7.

APOLLO 10 Preparation

The pace of exploration increases as activities are stepped up for the mission that will bring men and machines closer to the moon than ever before. The crew of Apollo 10 will perform almost every single task required for an actual moon landing, with the exception of landing on the moon itself. Apollo 10 is the last chance to iron out any "bugs" in the performance of the spacecraft, and the crew and technicians at NASA carefully follow the step-by-step mission plans set out for them. What they learn-and what they do- will make a lunar landing as safe and foolproof as possible. Six months before launch, the ascent stage of Lunar Module 4 (**1**) is prepared for lowering into the altitude chamber at Cape Kennedy's Manned Spacecraft Operations Building. Many tests are performed in this chamber, one of which is a docking test with the Command Module (**2**) which will carry the Apollo 10 astronauts on their planned mission. It is now February, 1969 (**3**), and inside High Bay 3 of the cavernous Vehicle Assembly Building, the three

stages of the massive Saturn V booster have been mated. Big as it is, it is going to become even taller when the three spacecraft modules, the adapter skirt, and the escape tower are attached to it. Just like Topsy, it keeps on growing (**4**), as the Apollo part of the vehicle is mated to the Saturn booster. Together, the completed vehicle is designated Apollo/Saturn 505. On top here is Command Module 106 *(Charlie Brown)* mated to its Service Module, which in turn is mated to the tapering Spacecraft Launch Adapter skirt. Inside the SLA skirt is the compartment in which Lunar Module 4 *(Snoopy)* is housed. *Charlie Brown* becomes more realistic (**5**) when you can measure its size in proportion to the men working on it during mating of the CSM to the adapter skirt. Actually, compared to the Saturn stages and the adapter, the Command Module is quite small and compact. It's almost hard to believe that three men will live and work inside it for eight days on their way to the moon and back. Rollout of A/S 505 (**6**) takes place early in the morning of March 11, 1969, as the completed rocket moves slowly away from the VAB. The Saturn V rocket, the Launch Umbilical Tower, and the Mobile Launch Platform are all carried atop the Crawler/Transporter, which is practically hidden here underneath its massive burden. Inching its way along the four miles of special roadbed prepared for it, the crawler arrives at Pad 39B ten hours later, late in the afternoon (**7**). This launch will be the first time Pad B is used.

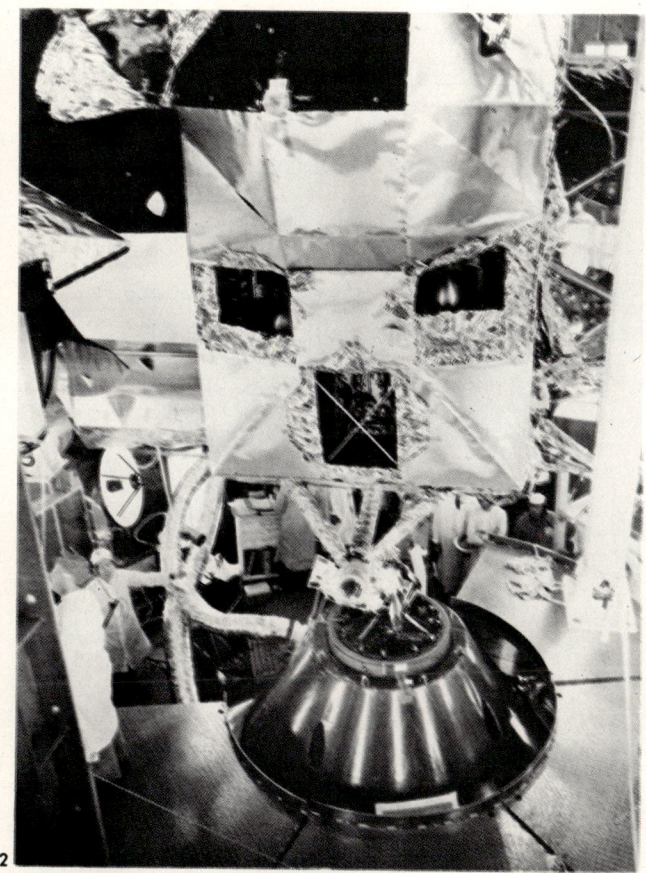

APOLLO 10
Pre-launch

Getting the spacecraft to the launch pad is just one of a series of thousands of steps that must be taken before the craft leaves the ground. (**1**) Dr. Hans F. Gruene, KSC Director of Launch Vehicle Operations (center), and Rocco A. Petrone, Director of Launch Operations (right), keep their eyes glued to television screens as they monitor the progress of the dry portion of Apollo 10's Countdown Demonstration Test on May 6, 12 days before launch. The successful test paves the way for the launch on May 18th. The big day arrives at last (**2**), and the astronauts breakfast with NASA officials before suiting up. Young (left foreground) seems to be grimacing over the prospect of yet more orange drink, while Stafford (second from right) explains a point to George Low, Manager of the Apollo Spacecraft Program. During the two-hour suiting-up procedure (**3**), the crew relaxes while technicians check out all their suit systems. All dressed up in his own space suit (**4**), Snoopy says "All systems are go!" in this banner being

shown to Stafford. Besides being the name for Apollo 10's Lunar Module, Snoopy is also the official symbol for NASA's Manned Flight Awareness Program, aimed at promoting high quality workmanship and performance. But where is Snoopy? Ahh . . . there he is! The famous dog himself turns out to greet the crew (**5**) as they leave the Operations Building for the launch pad. Holding the Snoopy doll is Jamye Flowers, secretary to astronaut Gordon Cooper, as Tom Stafford gets in a pat for some dog-gone good luck. Stafford again leads the way (**6**) as the crew debarks from the transfer van at Pad B. They enter the elevator in the Launch Umbilical Tower, which takes them 320 feet up to the white room. There they submit to last-minute ministrations (**7**) before entering the Command Module, the open hatch of which can be seen under Stafford's upraised arm. Just above his head hangs a large NASA mailing envelope, reading in bold and confident letters "Return Passage Guaranteed." While the attention of the press and the world is focused on the launch of Apollo 10, NASA's continuing plans quietly unfold. This (**8**) is Firing Room 1 in the Launch Control Center, which is conducting an unhurried checkout test of the Apollo 11 launch vehicle. Fully assembled, this rocket is still in the VAB, and due to be rolled out in only three days on May 21. These engineers are readying the next momentous step in space exploration even as their neighbors in Firing Room 2 noisily send Apollo 10 on its thunderous way.

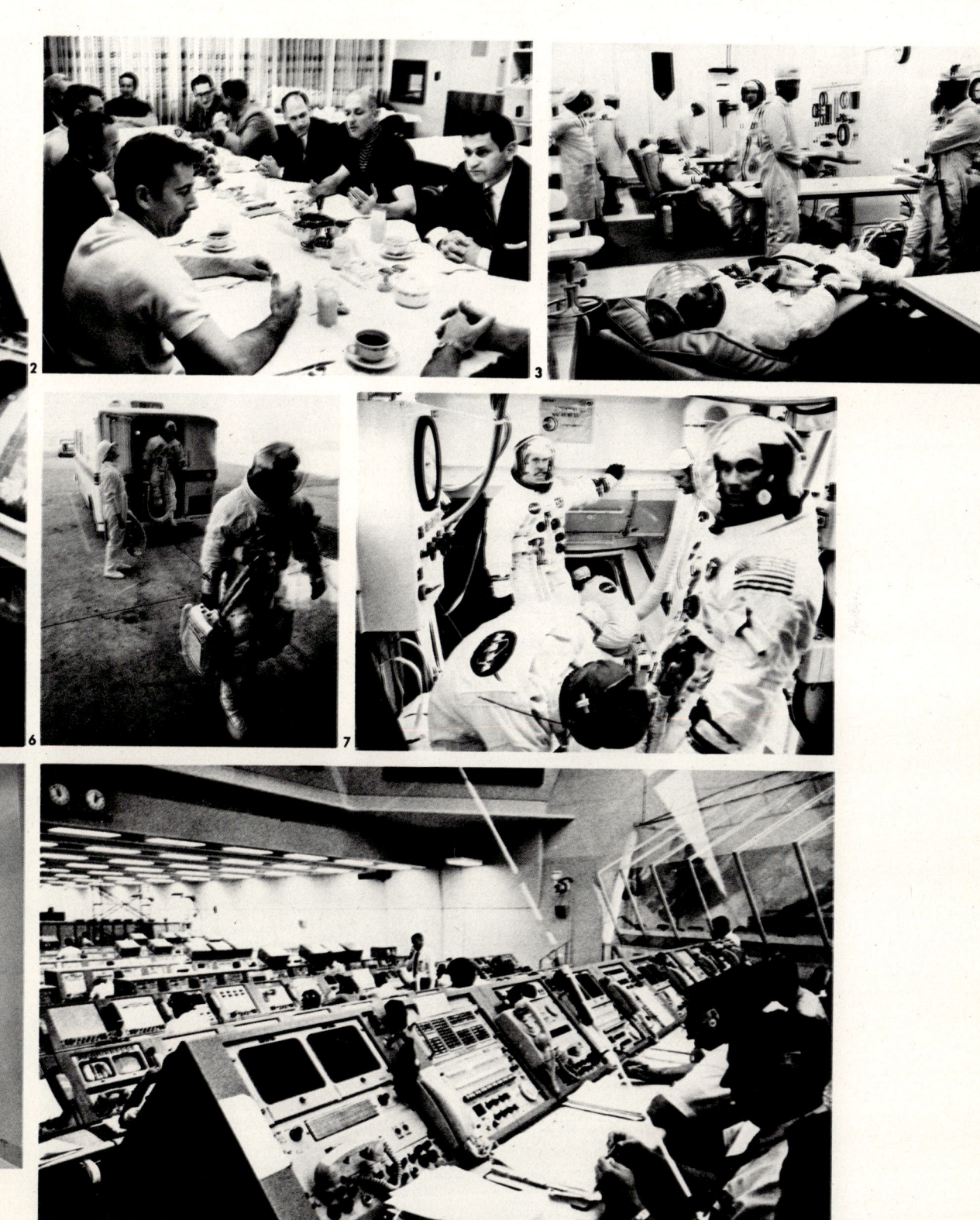

APOLLO 10 Launch

They're off! Gene Cernan's wife Barbara (**1**) tensely watches the billowing smoke clouds at ignition, while 6-year-old Tracy Cernan almost, but not quite, covers her eyes and ears. With the Cernans are Al Bishop (left), a family friend, and Father Eugene Cargill, the Cernan's clergyman. Mrs. Cernan points excitedly (**2**) as the mighty Saturn carrying her husband gains altitude, while Tracy and Father Cargill grin with delight. That's Tracy's *daddy* up there! Every bit as excited as the Cernans are former Vice-President Hubert Humphrey (**3**) and his wife Muriel, also watching the launch from the quest viewing area. To Mr. Humphrey's right is Mrs. Emil Mosbacher, wife of the chief of U.S. Protocol. In the second row are (left to right) Albert Siepert, Deputy Director of the Kennedy Space Center; King Baudouin of Belgium; Queen Fabiola of Belgium (behind Mr. Humphrey); and Mrs. Siepert. A camera's fish-eye lens captures Apollo 10's climb to the heavens from across a flower-strewn field (**4**), providing this contrast be-

tween nature and the space age, and what a glorious contrast it is! Ignition occurs at 12:45.5 PM EDT on Sunday, May 18, only a half-second later than scheduled. All the early parts of the flight proceed faultlessly in rapid sequence. The crew members are ecstatic over the power of their giant rocket. "What a ride!" exclaims Stafford. Eleven minutes and 53 seconds after lift-off, Apollo 10 has left Cape Kennedy far below and enters a near-perfect circular "parking" orbit, 100 miles high. The engine of the S-IVB third stage is then shut down while all systems are checked out. Midway through the second orbit, Mission Control gives the green light for trans-lunar injection, and the S-IVB's J-2 engine is restarted and fired for 5 minutes and 42 seconds. This raises Apollo 10's speed to 24,250 miles per hour—almost 7 miles per second, many times faster than any bullet—enough to escape earth's gravitational pull, and send it looping out on its three-day voyage to the moon. Now John Young begins a vital task. As CM pilot, he must retrieve the Lunar Module *Snoopy* from its compartment atop the S-IVB. He undocks the CSM, moves it a short distance away, turns a half-somersault, then reapproaches the third stage cautiously. Gently he eases the CM's probe into the LM's funnel-shaped docking collar, and the ten locking latches finally click into place. With the skill attained in long practice in simulators, Young now reverses the CM's thrusters, and pulls the LM free; *Snoopy* and *Charlie Brown* now fly together in space.

APOLLO 10
In-flight

The approach and docking with *Snoopy* is broadcast to Mission Control and the home audience in color, another space first. While Young handles the controls of *Charlie Brown* during the retrieval maneuvers, Gene Cernan operates the tiny 15-pound color television camera, a marvel of electronic miniaturization. This is the first of a series of TV transmissions to earth, and the camera was to work flawlessly throughout the mission. During the next colorcast Cernan turns the camera on earth, and we could see for the first time a living view of what our hauntingly-beautiful planet looks like from deep space. Scientists and poets alike have coined the phrase "Spaceship Earth" for our tiny multi-colored self-sufficient world, so small and fragile-looking from this distance. The cloud-covered globe can be seen on Mission Control's wall screen (**1**) as it is being transmitted from the speeding spaceship, 12,000 miles away. The astronauts are in high spirits on the outward flight, sending back weather reports and playing music

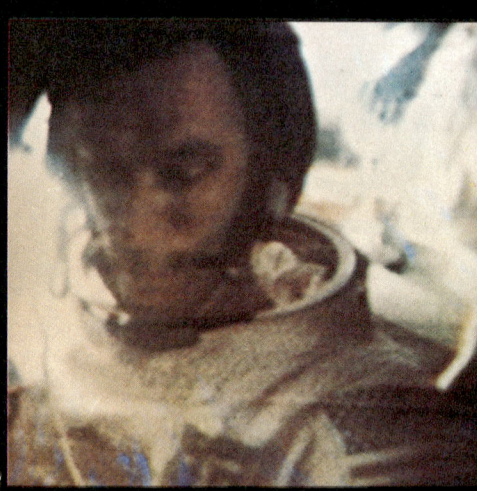

ing "In My Beautiful Balloon" and—especially appropriate—Frank Sinatra's rendition of "Fly Me To The Moon." LM pilot Cernan is hard at work (**2**) in this picture made from the third TV transmission, while the spacecraft was about 36,000 miles from earth. No, Gene Cernan isn't lost and looking at a celestial roadmap here (**3**); that's his lunch. He's mixing freeze-dried food with water in a plastic bag, and assures us it will taste better than it looks. We'll take your word for it,
done. There's a special male comfort on this trip; for the first time, astronauts can shave in flight, thanks to a new shaving cream that holds their shorn whiskers. During their fourth telecast (**4**), Stafford (left) gags it up for the home audience: he's holding an upside-down and weightless Young with one hand. Now rightside-up, Young (**5**) displays his own gag—and considerable artistic ability—in this portrait of Charlie Brown in NASA coveralls. Drawing them ever onward, the still-mysterious moon (**6**) beckons
alluringly from the black depths of space as the astronauts cross over into its field of gravity, 205,000 miles from earth. In this view, the lunar north pole is slightly to the left of top. The prominent dark "seas" are the Sea of Serenity (upper left), Sea of Tranquility (center), Sea of Fertility (center right), and the Sea of Crises (upper right), all long since bone-dry. Apollo 10's speed has dropped to 2000 miles per hour, but gradually rises again to over 5000 mph during the one-day "downhill coast" to lunar orbit.

APOLLO 10
In-flight

Apollo 10 is scheduled to enter lunar orbit on its third day of flight out from earth. As Apollo 10 comes close to the moon it will be flying at speeds of about 5500 miles per hour. The next step is for the crew to prepare for the Lunar Orbit Insertion (LOI) burn, which will drop the spacecraft into a long, oval orbit around the moon. The ship's speed will have to be reduced considerably as it enters lunar orbit at an altitude of about 113 miles above the moon. The slowing down, or braking, is always done behind the moon, on the dark side that is never seen from earth. The critical LOI burn and entrance into lunar orbit is completed successfully, and after Apollo 10 swings out from behind the moon the crew informs Houston that they have arrived. Cameras aboard the spacecraft start clicking away, recording the moon and sky-scapes, as well as the route for the scheduled landing next July of Apollo 11. The astronauts also prepare for the separation of the LM *Snoopy* from *Charlie Brown*. Through the rendez-

can be seen the thin polished skin of the top of the docked LM. In the upper right is a portion of the LM's narrow docking window, on which are visual aid markings to help the pilot during docking maneuvers. Undocked and flying free (**2**), *Snoopy* is snapped by Young as the LM temporarily trails alongside the CM. Inside, Stafford and Cernan prepare to lower their spidery craft down to an altitude of 9 miles, so that they can get a really close look at the surface.

Nine miles may still sound quite high, but that's about 50,000 feet, lower than many military jet planes fly above the earth every day. From his vantage point in orbit, John Young sees this nameless crater (**3**), designated No. 302 by the International Astronomical Union. The IAU is an international group of astronomers whose job (among others) is to catalogue all lunar features, including those with no formal name. Though on the far side of the moon, IAU 302 exhibits the terracing and central peaks commonly found in large craters on the near side. Schmidt Crater, however (**4**), on the western edge of the Sea of Tranquillity, is a sharp-edged young crater. Six miles in diameter, its western wall reflects the rays of the rising sun in a new lunar day. This (**5**) is the western region of the Sea of Tranquillity, the area planned for the landing of Apollo 11. The exact spot, in fact, is at the top center of the picture, to the right of the small crater Moltke B, with a landing approach over one of the smoothest areas of the moon.

APOLLO 10
In-flight

As Stafford and Cernan in *Snoopy* zip around the moon at an altitude of only 9 miles, John Young, in *Charlie Brown*, continues his own orbital pattern, some 60 miles high. Stafford and Cernan have a marvelous vantage point; from their lower altitude they can clearly see, identify and explore the lunar surface accurately. Houston keeps asking them to check on various features of the moon's terrain, and while their comments are somewhat unscientific as far as geological nomenclature is concerned, they cannot be faulted as to the accuracy of their descriptions. No one has ever been as close to the moon as they are, so the information they are sending back to Houston is of the utmost value. On closer inspection, it turns out that the future landing site of Apollo 11 is not so smooth as it originally appeared. The astronauts state their conviction, however, that if the LM to be used during the Apollo 11 landing has enough hover time, there should not be any problem in landing. In addition to flying the CSM, Young keeps

in touch with the LM, reports to Houston, and takes numerous pictures. He will soon have to come down closer to the moon to meet for the rendezvous and docking with the LM, and it *must* be successful in order to prevent Stafford and Cernan from becoming the first human lunar satellites. Simulating a real blast-off from the moon, the LM crewmen shed the lower descent stage of their craft, and climb upwards using their ascent stage engine. At rendezvous, Stafford and Cernan have this view (**1**) of the glittering CSM, while Young sees *Snoopy* this way (**2**) as the two spacecraft maneuver for docking. Together once again, the crew bids goodbye to trusty *Snoopy*, and with a remote-controlled blast from its engine, sends the ascent stage of the Lunar Module off to eternally roam the solar system. The always-spellbinding spectacle of earthrise (**3**) now holds their attention, as they look across the barren crater-pocked fields of the moon to their home—our home—a bright, glowing marble a quarter of a million miles away. That is their destination; for now the Service Module's big Aerojet General engine is fired up again, and *Charlie Brown* swings out of lunar orbit, headed home. During the 3-day return journey, Stafford and Young (**4**) demonstrate how little room there really is in the Command Module, dominated here by the controls and switches for the navigation computer. A reluctant sacrifice for science and neatness (**5**); Young ruefully shaves off the luxuriant Edwardian-style beard he has been pampering for a week.

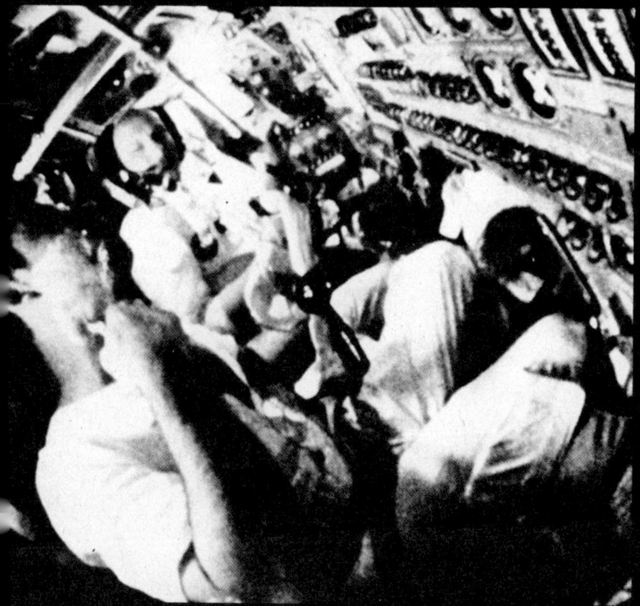

APOLLO 10
Splashdown

The journey towards home is smooth and uneventful, and the crew settles down into the almost routine tasks of returning to earth and splashdown. The disconnecting sequence from the Service Module goes without a problem, and the Command Module, now on its own internal power, survives the blazing heat of re-entry as designed. It is dawn in the South Pacific, that romantic and languorous part of the world extolled in song and story, the setting for some of the greatest tales of James A. Michener, W. Somerset Maugham, Norman Mailer, and the music of Rodgers and Hammerstein as well as haunting native strains. But its idyllic skies and placid seas never looked more welcome to anyone than the trio in *Charlie Brown* (**1**), as the Command Module floats gently down against the brightening glow in the eastern sky. The round trip has taken 192.03 hours—exactly eight days—from lift-off to this splashdown on May 26, only four miles from the prime recovery ship, the carrier U.S.S. *Princeton*. The astronauts are

greeted by the traditional frogmen and Sikorsky helicopter (**2**), as they clamber out of the CM into the life raft. The sun is now brightly shining as the helicopter is positioned on the *Princeton's* flight deck (**3**), and the ship's crew lines up to cheer the astronauts as they descent onto the waiting red carpet. Everyone aboard gathers around (**4**) as the newly-returned moonmen prepare to address the crew, to whom this is a high point after weeks of preparation. Flags wave gently in the breeze, Navy officers wait attentively, and the speeches and welcome-home ceremonies are about to start. After a prayer of thanks by the ship's chaplain, the astronauts briefly recount the highlights of their trip, concluding with, as always, "It's good to be back." There is probably a ship's band somewhere in the background, but certainly no sign of lovely native girls laden with flowers—even though the ship is near Pago-Pago. Oh well; maybe next time . . . Crew members and NASA officials stand by as *Charlie Brown* is brought aboard the flight deck after recovery (**5**), looking as charred and tattered as Command Modules generally do after undergoing the searing heat of re-entry. The capsule is so *small* —especially when compared to the enormity of the mission! Relaxing in borrowed bathrobes, Stafford, Cernan, and Young (left to right, **6**) chat on ship-to-shore phones with President Nixon, and assure him that not only are they hale and hearty, but that the way has now been cleared for the upcoming manned landing in two months.

The "music of the spheres" influences all of us, but none so directly as those involved in space flight. Nothing remains motionless in our solar system, which itself is moving. The planets revolve about the sun; satellites, like our moon, revolve about both planet and sun, while all rotate on their own axes. How do men manage to travel from one selected point to another in the solar system? Space travel is indeed complex, but is firmly based on principles of astronomy and navigation. The total of all these variables is called the *launch window*. Essentially it is an imaginary hole, or "window," in space through which a rocket is fired at a given time. To determine the proper window, it is first necessary to define the basic set of requirements for the mission. For a lunar landing mission (**1**), there are many such requirements. For good visual tracking, the launch site at Cape Canaveral must be in daylight. For safety reasons, the launch direction from Canaveral is restricted to an arc of 72 to 100 degrees east of north. All of this down-range

THE LAUNCH WINDOW
How NASA hits a moving target from a moving platform

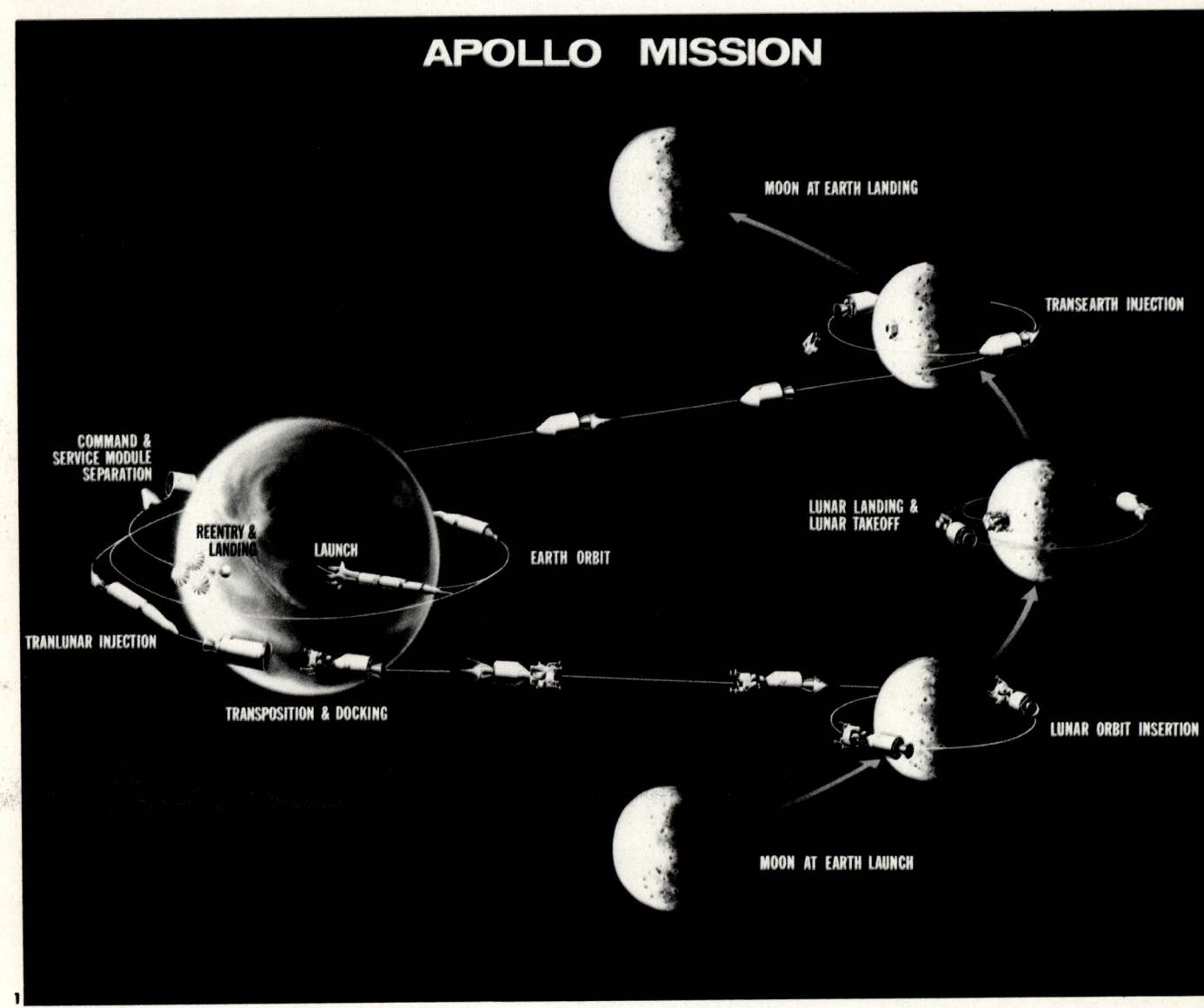

area must also be in daylight, so that if the rocket should fail to reach orbit and fall into the sea, the astronauts can be seen and recovered quickly. This requirement alone allows only one 4-hour launch period each day. The translunar injection burn has to occur over the Pacific Ocean, due to the position of the key tracking facilities at Goldstone, California. A very important consideration is the sun's elevation at the lunar landing site. Low sun elevation angles, from four to thirteen degrees above the lunar horizon, are required to create shadows, which aid the crew during landing. In other words, it must be early dawn or late evening in a lunar day. Since it takes three days to reach the moon, navigators must calculate backwards from the proper landing time to find the proper launch time. Goldstone must be in direct communication during the lunar landing phase; this occurs for 12 hours a day. The last requirement is for a daylight earth splashdown. This eliminates most launch opportunities for landing sites on the far eastern part of the near side of the moon. An interesting situation is that the splashdown will always occur at a point near the "lunar antipode" from the injection burn (**2,3**). "Antipode" is defined as "opposite part of the earth"; thus lunar antipode refers to a point on earth exactly opposite that of the transearth burn behind the moon. Sunlight conditions are illustrated for a eastern lunar landing site (**2**) and for a central site (**3**). Computers assemble all this data, and then select the best launching times.

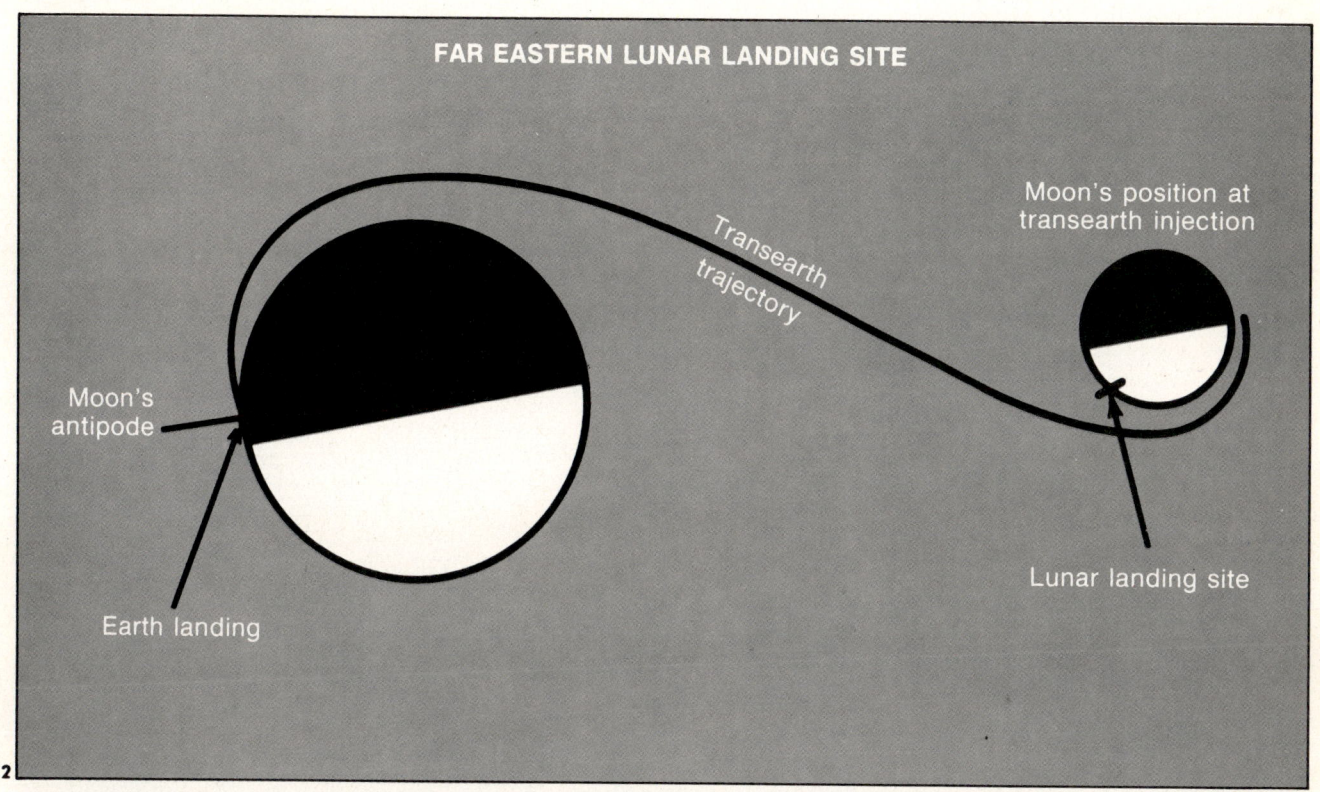

FAR EASTERN LUNAR LANDING SITE

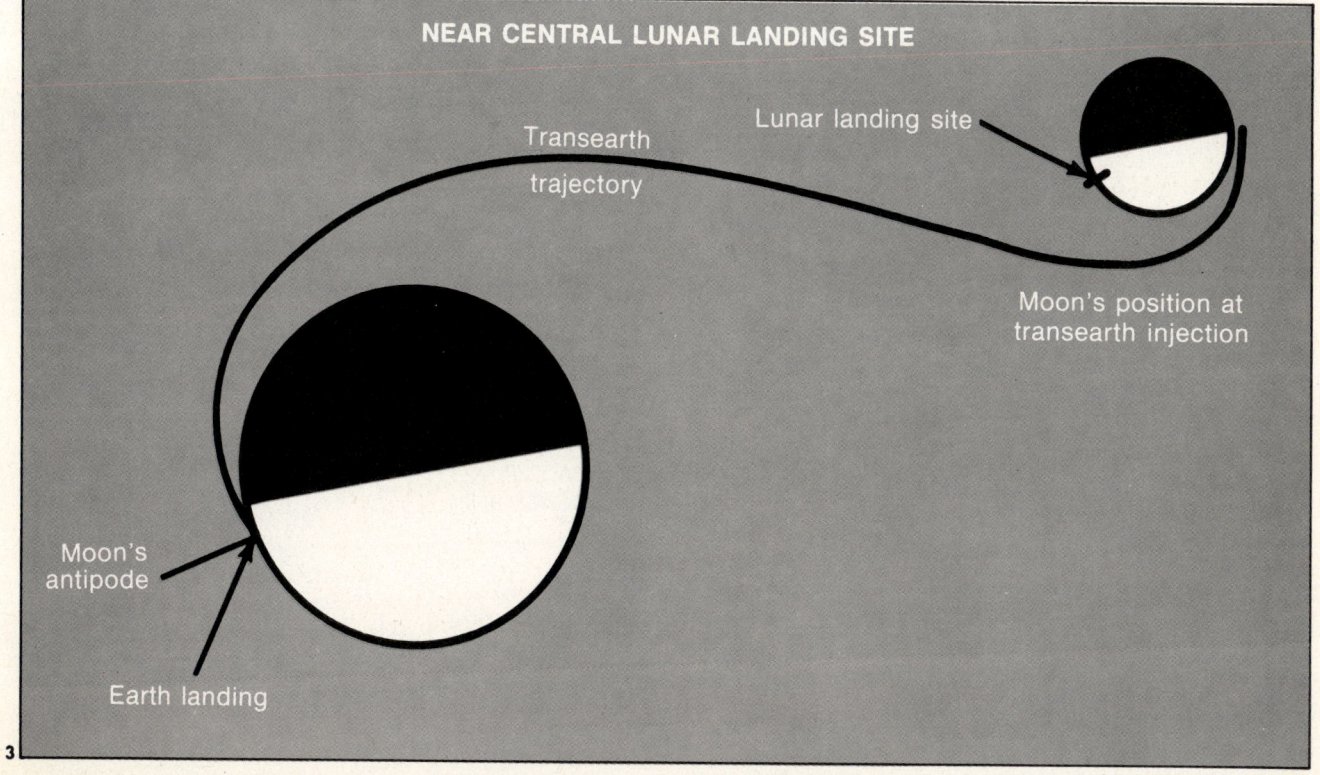

NEAR CENTRAL LUNAR LANDING SITE

Space is a hostile environment for man, who lives at the bottom of an immense ocean of air which earth's gravity conveniently holds in for him. This air, or atmosphere, contains the oxygen which man must breathe. One-fifth (21%) of air is oxygen; almost all of the rest is nitrogen. But air provides more than just oxygen; it has *pressure* (14.7 pounds per square inch at sea level), which forces the air into lungs and prevents the water-based human fluids from evaporating too rapidly or even boiling; it has *density*, which destroys or minimizes high-speed space particles and harmful radiation; it has *humidity* —water vapor—without which man's tissues would dehydrate; and its *motion* controls temperature within the relatively narrow range of human tolerance, as well as carrying away waste products. In space there is no gravity and no air. Tests conducted during the 1950's showed that man could probably function quite well for extended periods in zero gravity, but he must find a substitute for the atmosphere or else take it with him.

SUIT-UP FOR SPACE
A new coat of armor shields man against a hostile environment

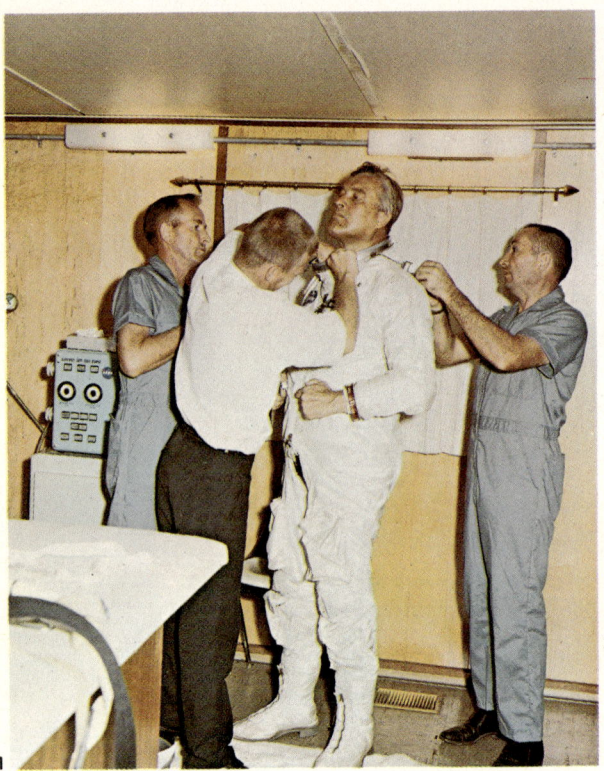

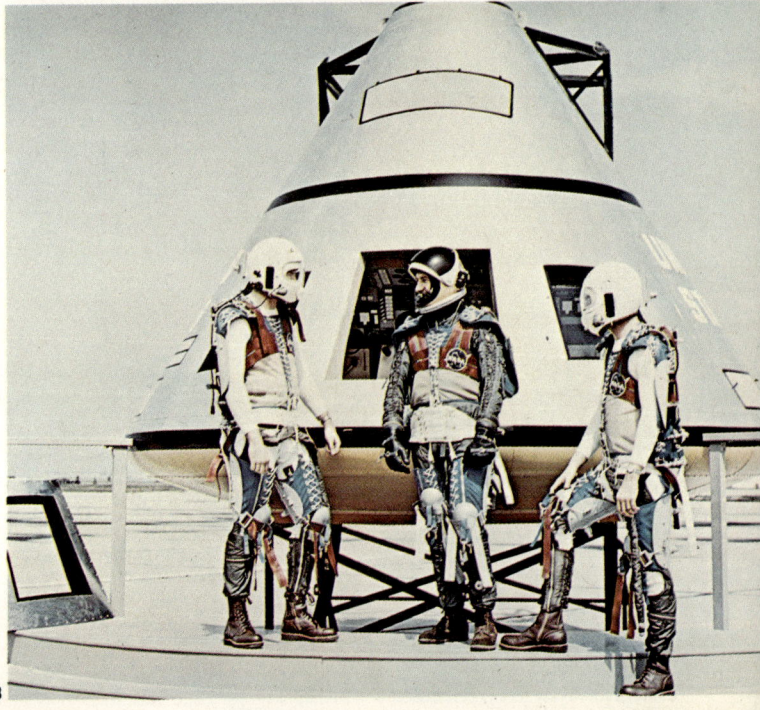

That is the function of the space suit. During the 1930's aircraft performance began to penetrate into the stratosphere, where conditions are not far removed from space. Some sort of protective suit became necessary, and the first great pioneer in this field was none other than Wiley Post, the famous one-eyed aviator who tragically perished with Will Rogers in 1935. Post realized that breathing pure oxygen under low pressure was equivalent to breathing air at much higher pressure, and he devised several successive suits designed to hold oxygen at a pressure of only three psi. He was plagued with the problems which beset later designers; either the suit was so stiff when inflated its occupant could hardly move, or it leaked or ruptured when made flexible. His most successful suit is now on display in the Smithsonian Institution. Project Mercury suits (**3,** center, and **4**) were made by B.F. Goodrich—which made Post's suits—and were adaptations of high-altitude pressure suits. Gemini suits (**5,6,7**) were made by the David Clark Company of Worcester, Mass., another pioneering firm which has been making high-altitude pressure suits since 1939. The extravehicular version (**7**), used for spacewalks, shows the reflective gold-plated visor later used in the Apollo program. In this rendering of the LM lunar landing (**2**), the artist depicts Gemini suits, as the latest Apollo suits had not yet been developed. Even Wernher von Braun was fitted with a suit (**1**), as he wanted to sample first-hand the rigors of astronaut training.

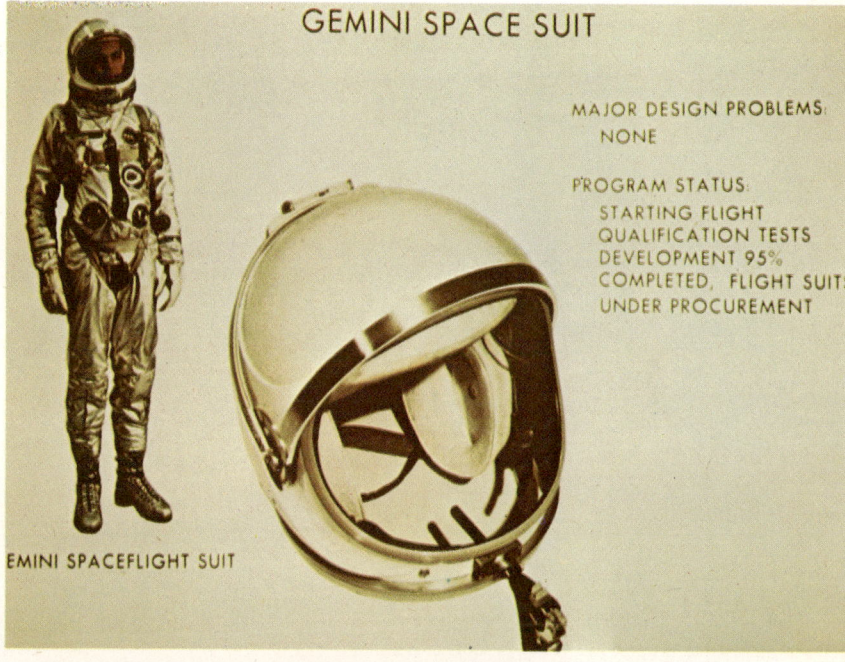

MAN IN SPACE VOL. 4/115

SUIT-UP FOR SPACE

The major goal of Project Mercury was to ascertain whether a human being could live and function properly after 24 hours in space flight. The astronaut was little more than a passenger in the capsule, which was operated almost exclusively from the ground. Since the capsule was pressurized and not meant to be opened in flight, the Mercury suit was not a true space suit; it was an emergency suit always worn uninflated. If the spacecraft lost cabin pressure, the suit was designed to automatically inflate to 3½ pounds per square inch, and Mercury Control would then bring the capsule down as quickly as possible. Costing $10,000 each, it was an adaptation of a Navy high-altitude suit, with an aluminized outer layer, an Air Force helmet, and a special inner ventilated rubber bladder. When inflated, it offered only limited mobility, which was adequate. However, at no time in any Mercury flight did a capsule develop a leak. Project Mercury reached its design objectives on the last flight, when Gordon Cooper in *Faith*

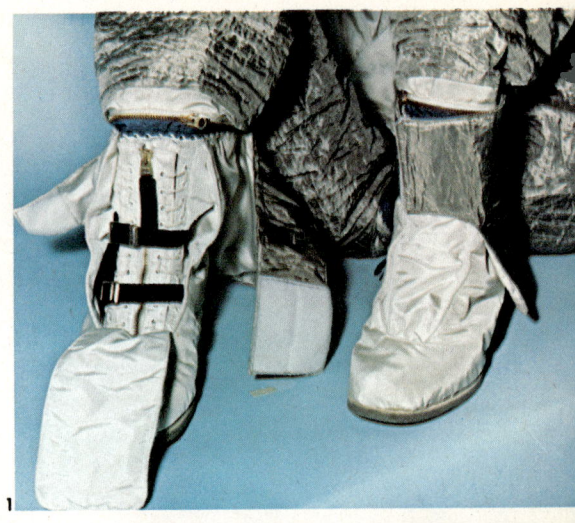

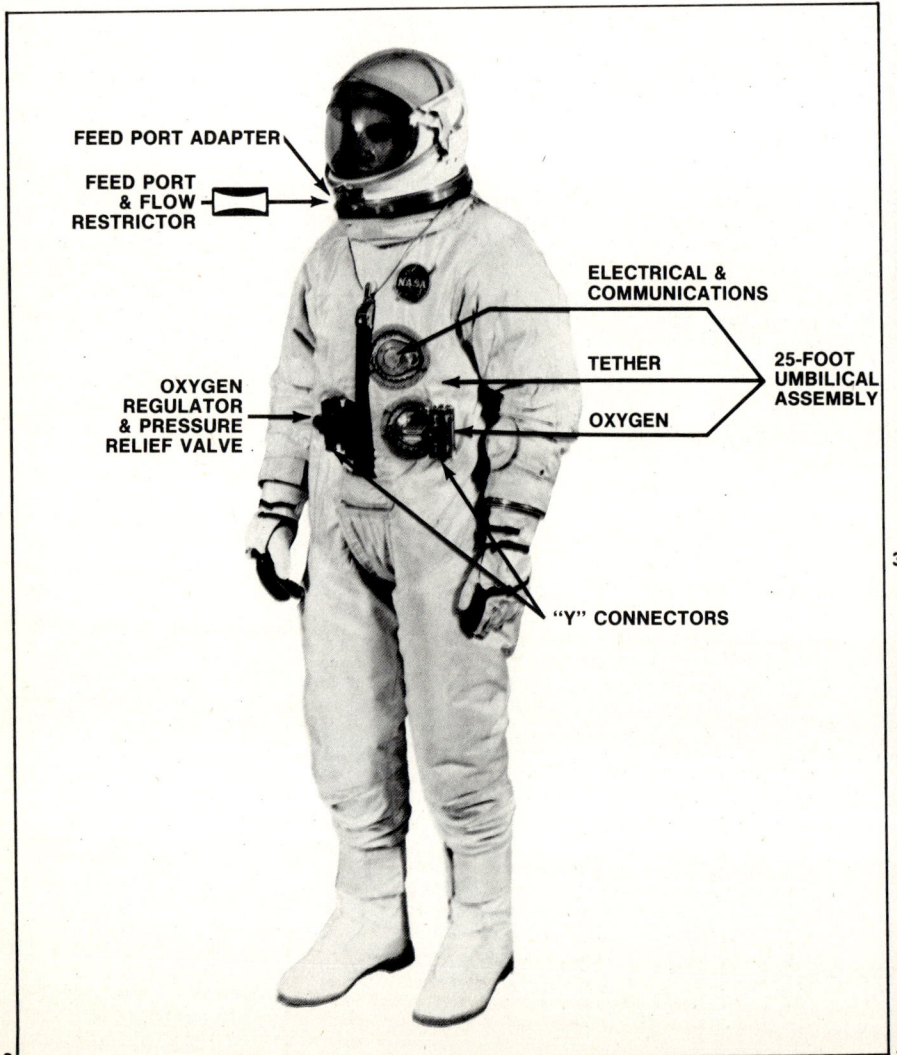

7 made 22 orbits, remaining in space for 34 hours and 19 minutes. In Project Gemini, however, astronauts were intended to perform useful work, pilot the maneuverable spacecraft, venture out into space unprotected by a pressurized cabin, and stay in orbit for as long as two weeks. The David Clark Company developed a new suit construction method which allowed much greater mobility when inflated. Called the link-net method, it was similar in principle to the chain mail armor worn by medieval knights. Elaborate lace-up boots (**1**) were part of the Gemini suit. The extravehicular version (**2**), which cost $80,000, was connected to the spacecraft by a 25-foot umbilical line which contained a tether, oxygen hose, and electrical lines, somewhat like a deep-sea diver. Apollo suits are made by the International Latex Corp. of Dover, Delaware, the same firm which manufactures "Playtex" undergarments. Each suit is custom tailored for a specific astronaut, and his body measurements result in patterns like these (**3**); approximately 1200 are required for each suit. The bellows-like joints for elbows and knees are made by wrapping nylon mesh around a mold (**4**), then dipping it into liquid neoprene synthetic rubber (**5**). The worker is blowing through a small tube to force the rubber deep into the convolutions. The pressure gloves are made with measurements taken from plaster casts of each astronaut's hands; these (**6**) are those of Ron Evans. Special molds are then fabricated, which are dipped in neoprene (**7**) to form the gloves.

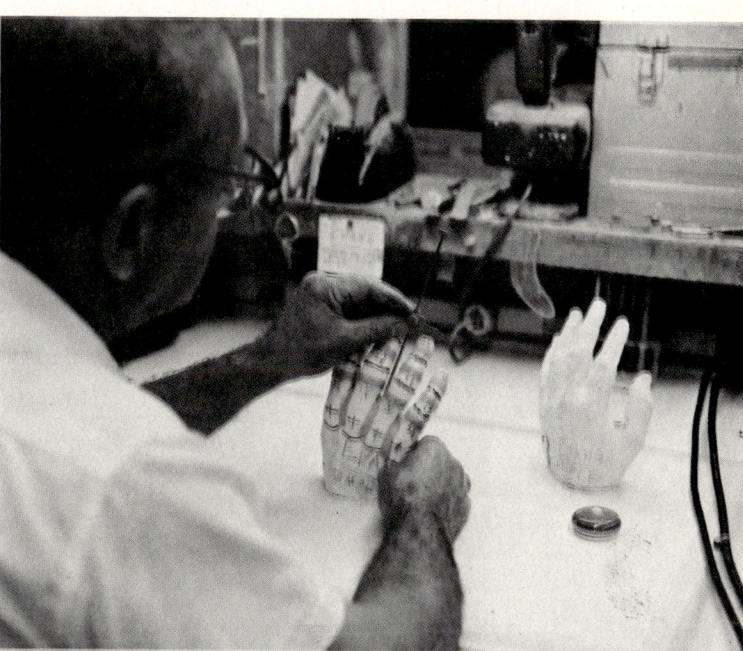

COURTESY OF ILC

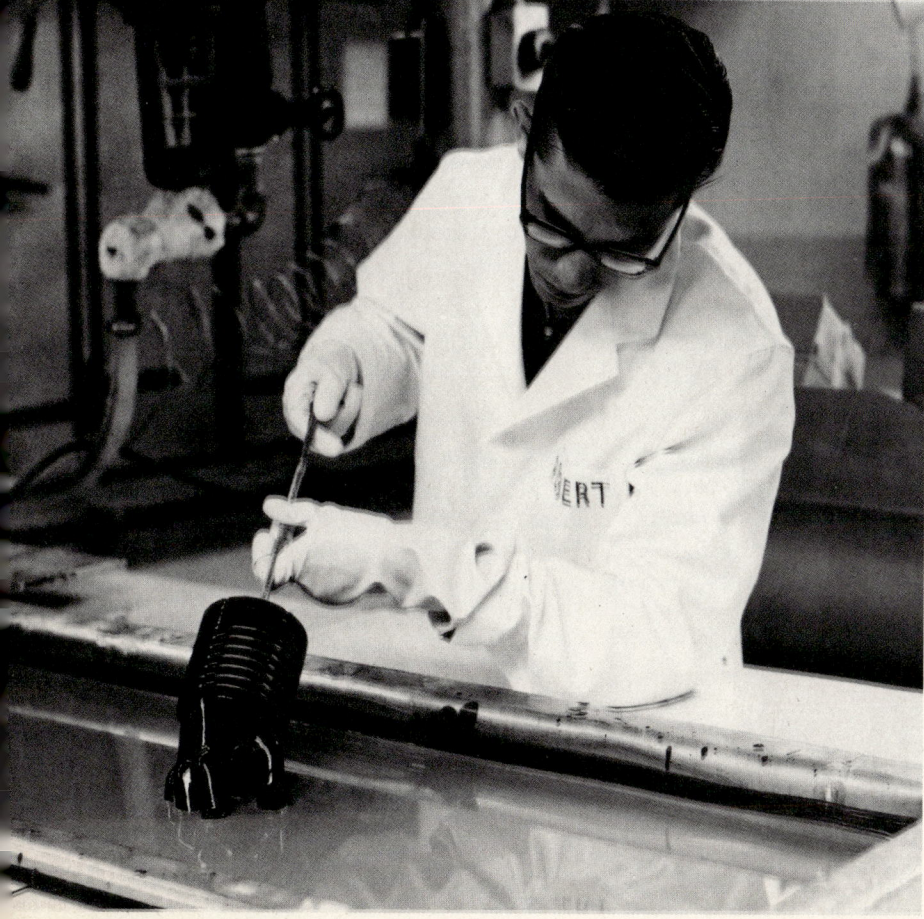

SUIT-UP FOR SPACE

The Apollo boots are also custom-made to fit individual astronauts. Of quite different construction than the laced designs used in Projects Mercury and Gemini, the Apollo boot is a full-pressure type which attaches to the suit with a sealed zipper, and incorporates a flex joint at the ankle. It is easier to put on and much more comfortable than the earlier designs. Here (**1**) a finished boot is pressure-checked at the International Latex plant. Rigorous inspection (**2**) accompanies every production step. On difficult parts like these boots and gloves, rejection rates sometimes reach 99%, which means only 1 out of every 100 articles is finally acceptable to the government. The inspector at right is examining the special lunar glove, which fits over the normal pressure glove and will be worn only by moonwalking astronauts. A close examination of the wrist joints of the pressure gloves piled at right reveals that the joint segments are indented at irregular intervals. This is a deliberate design feature; it prevents the segments from all bending at

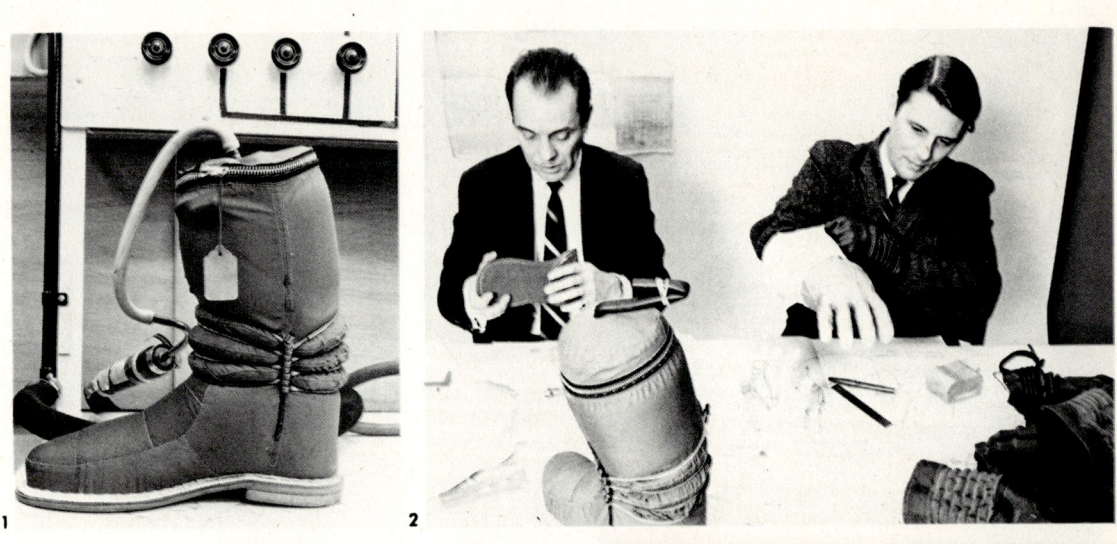

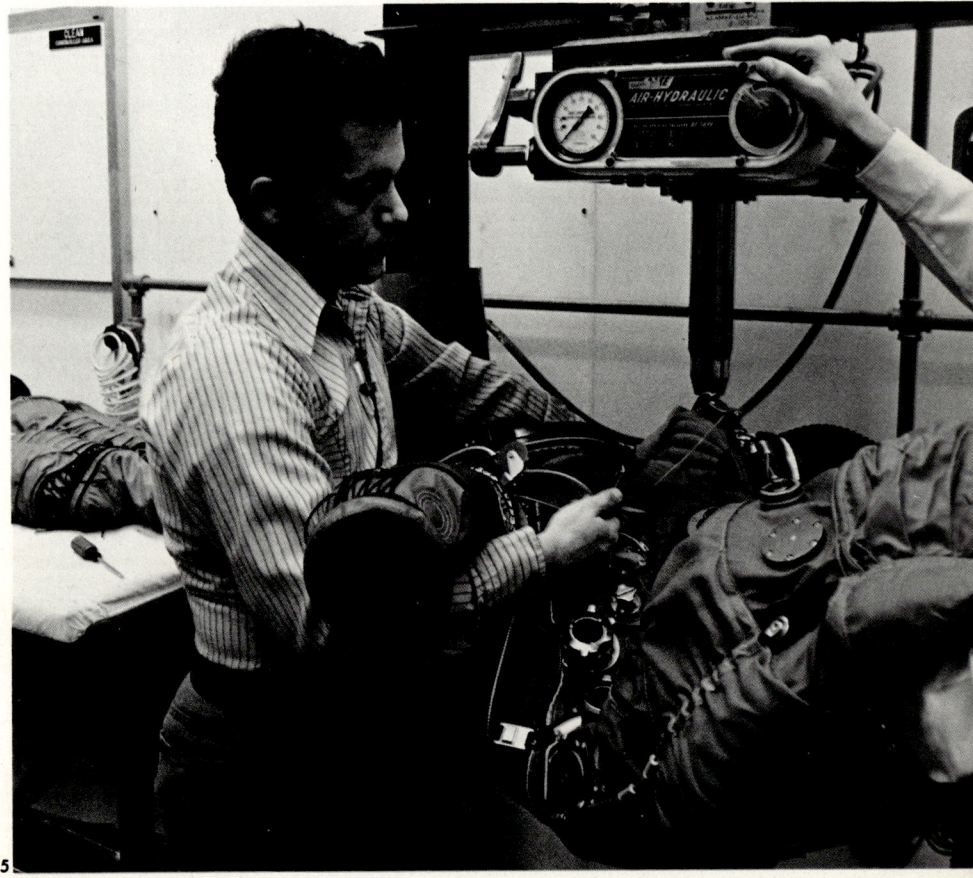

the same point when the wrist is bent, which would trap the air inside and raise its pressure. Suit production (**3**) involves the sewing together of many different layers of fabric. The intravehicular suit has an inner layer of fire-resistant Nomex nylon, followed by the pressure bladder of neoprene-coated nylon, then by another layer of nylon which prevents the bladder from ballooning when inflated, another layer of Nomex, and finally two top layers of Beta glass cloth. Beta fiber is made by the Owens-Corning Corporation, and is an extremely thin fiber made of pure glass. When closely woven into a fabric by J.P. Stevens & Co., these fibers form a lightweight cloth that is both fireproof and an efficient heat insulator. For abrasion resistance, the glass fibers in the space suit are coated with Teflon, the same slippery Du Pont material found in cookingware. This lady is working with one of the several additional layers of special fabric that form micro-meteroid protection for the extravehicular suit, worn by Apollo lunar astronauts. In the final stages of production, all the special hardware is attached (**4**), and the suit pressure-tested (**5**). Trial fittings (**6**) are done on persons who have the same general physique as the intended astronaut, with the final fittings on the astronaut himself. Mobility and operations checks (**7**) are then made. It is fascinating that at the very center of the Apollo program, its image dominated by impersonal computers and giant rockets, that so much old-fashioned handwork is required to properly equip man.

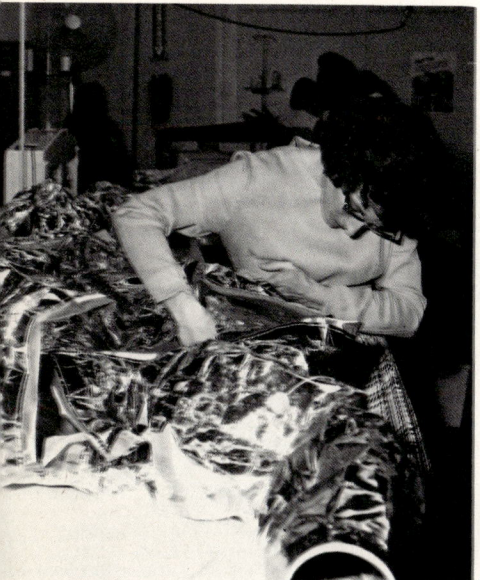
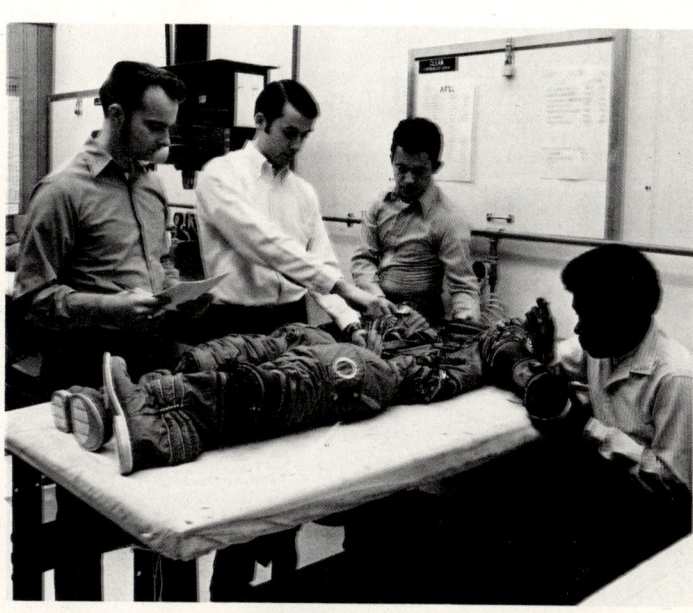

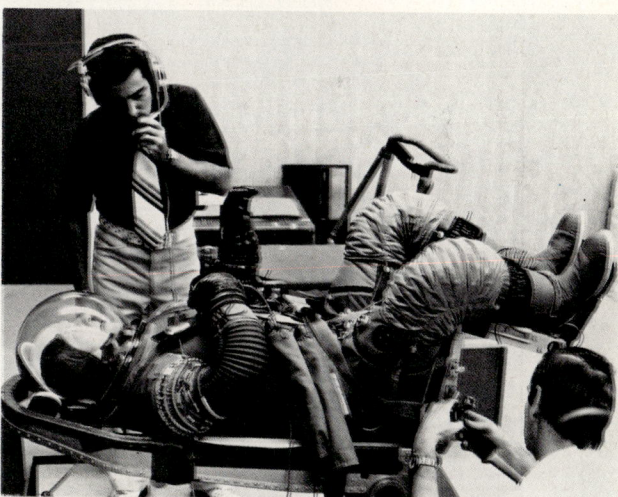

COURTESY OF ILC

SUIT-UP FOR SPACE

By 1968 the extravehicular suit, modified following the tragic Apollo 1 fire, looked like this (**1**). Its color had changed from light blue to white, it incorporated the new Beta-glass outer layers, and the formerly separate thermal/micrometeoroid cover garment was combined with the pressure suit. A waist joint will be added for Apollo 15 and subsequent missions, so that astronauts may sit in the Lunar Rover moon car. A major advance of Apollo suits is the one-piece fishbowl helmet (**2**). Previous types were worn directly on the astronaut's head; this helmet is connected to the suit, which relieves him of having to move the helmet's weight every time he moves his head. It is made of an incredibly tough General Electric synthetic material called Lexan; you couldn't break it with a baseball bat. On the moon it is covered with a cap, visors, and side screens (**4**). The various parts of the moon-walking suit are denoted here (**3**). Most vital is the Portable Life Support System (PLSS) backpack, made by the Hamilton Standard

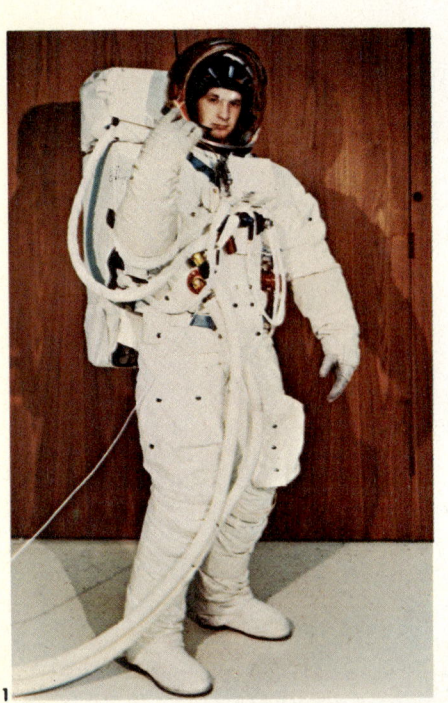

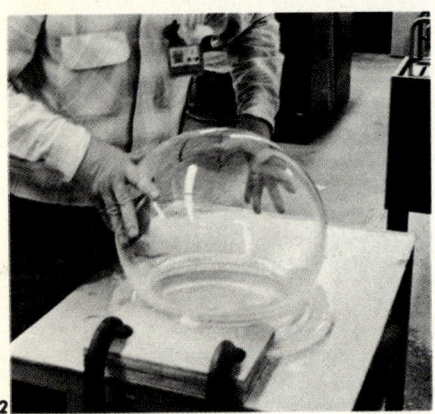

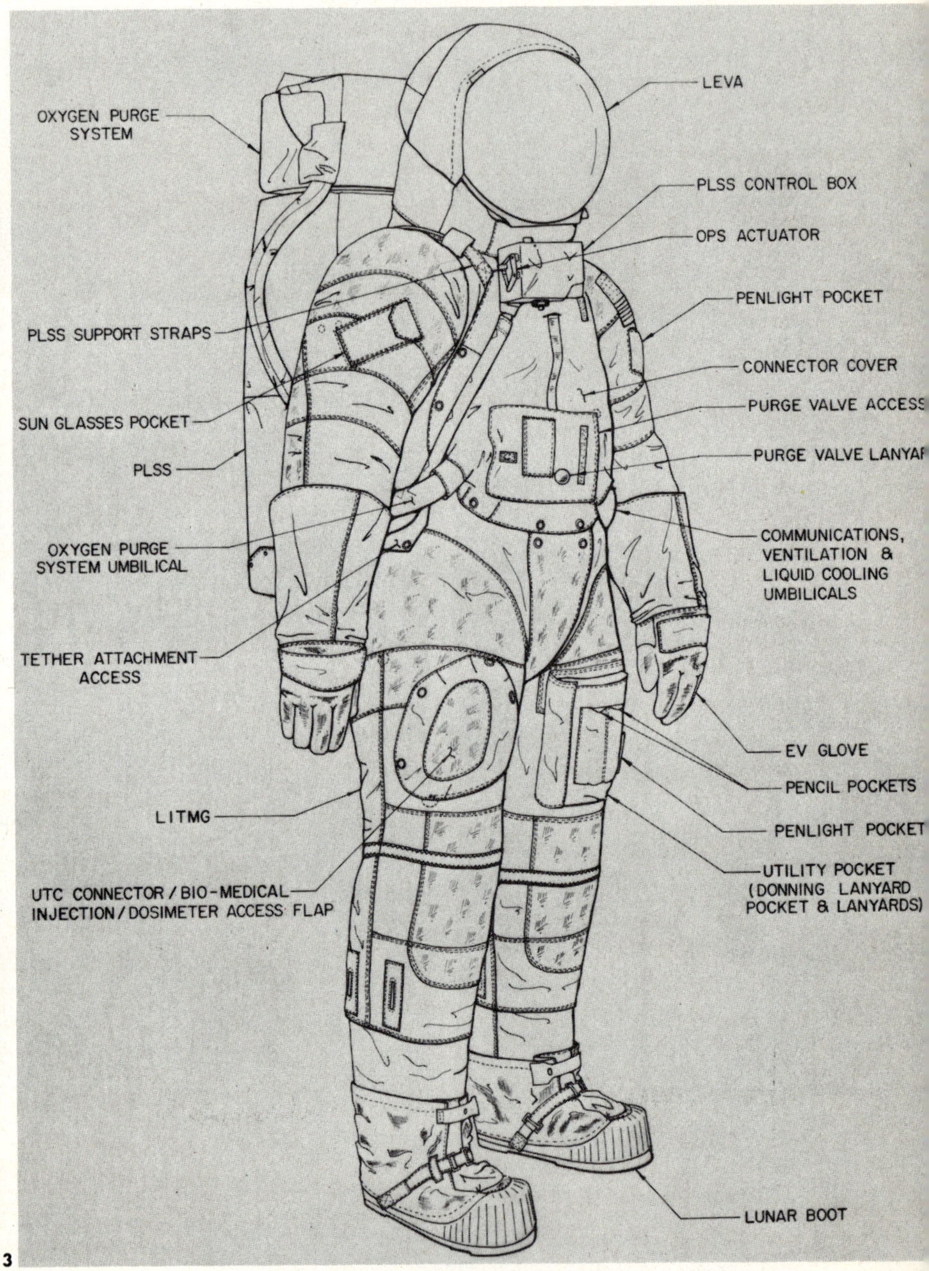

Division of United Aircraft Corporation. This is a unit which has three main functions: it recirculates oxygen, cooling water, and provides communication. The inside view (5) shows the part worn next to the astronaut's back; (6) is an outside view with the protective fiberglass shell removed. The *oxygen system* routs oxygen from the suit through an activated charcoal filter, which removes body odors, then through a lithium hydroxide canister, which removes carbon dioxide. It is then cooled and dried in a sublimator and returned to the suit. Any additional oxygen needed to maintain suit pressure of 3.7 psi is picked up from the primary tank, which holds 1.03 pounds of the precious gas. Should this system fail, a 30-minute emergency supply is carried in a separate unit on top of the PLSS. The *water system* contains 1½ gallons, constantly circulated by a pump through a network of plastic tubes next to the astronaut's skin. This water is cooled by the sublimator; the astronaut can vary water temperature with a 3-position diverter valve. The *communications system* uses the LM as a relay station. The astronaut can receive and transmit by voice, and seven channels of telemetry are constantly broadcast to Mission Control. PLSS power is a 17-volt silver-zinc battery, and a control panel is on the space suit. The PLSS has a 4-hour operating time, and can be resupplied three times with material carried in the LM. It weighs 84 pounds, plus 40 pounds for the emergency oxygen system; on the moon this is only 21 pounds.

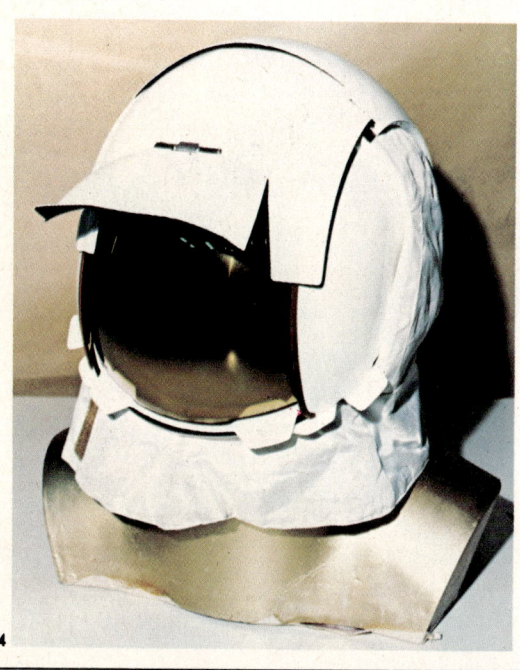

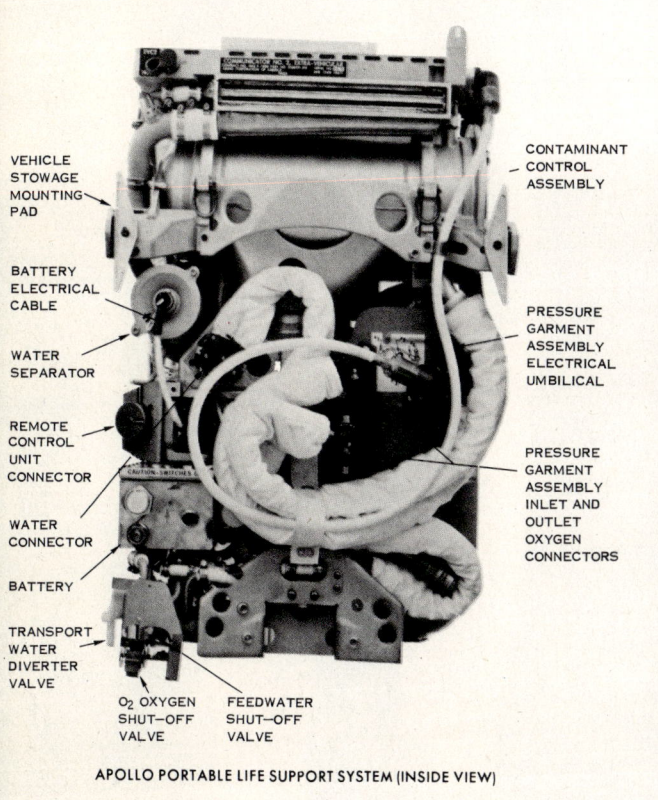

APOLLO PORTABLE LIFE SUPPORT SYSTEM (INSIDE VIEW)

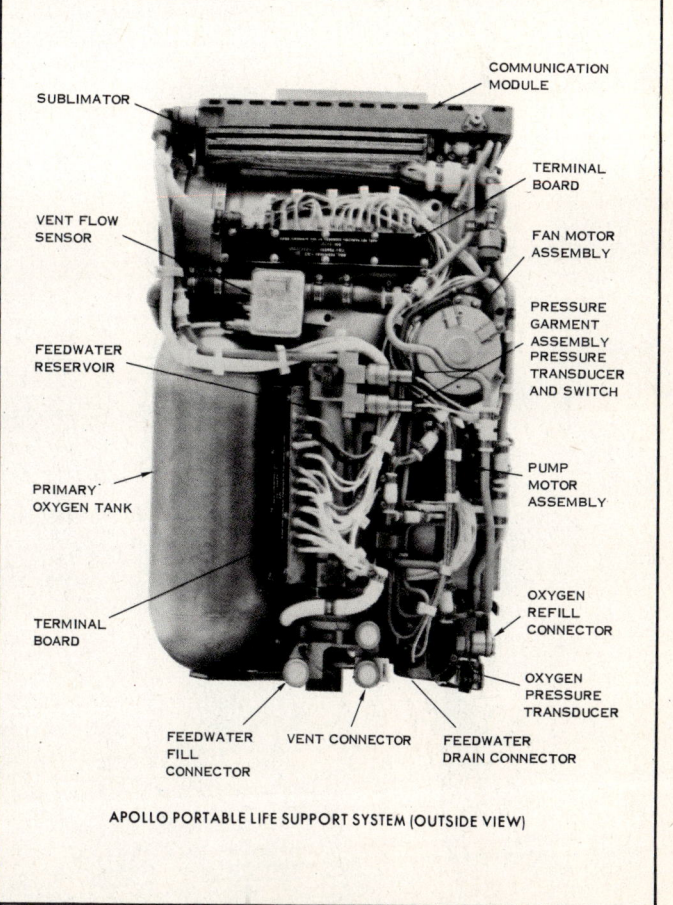

APOLLO PORTABLE LIFE SUPPORT SYSTEM (OUTSIDE VIEW)

FOR SPACE
INTRAVEHICULAR

While inside the protection of a Command Module or Lunar Module, an Apollo astronaut wears clothing which offers freedom of movement. His "long johns" are called the *constant-wear garment,* a full-length underwear that is his basic foundation garment. This covers the *bio-instrumentation harness,* fitted to him before the flight. The harness contains sensors which monitor his body functions, such as heartbeat and respiration rate, which are continuously relayed to Mission Control; it also has connections for the spacecraft's waste management system. If the module is pressurized, providing a "shirt-sleeve" environment, the astronaut simply wears his flight coveralls over the constant-wear garment. He also wears his *communications carrier,* a cloth cap containing a tiny microphone and receiver. If the cabin is to be depressurized, the astronaut removes the coveralls and dons his *intravehicular cover,* a lightweight space suit. This is a flexible pressure suit with no more than six layers of cloth, including an outer layer of fire-resistant glass fiber Beta cloth. Part of the suit are the helmet, boots, and pressure gloves, which are all sealed to prevent any leakage. Three connections on the suit (oxygen inlet, oxygen outlet, and electrical) hook up to the spacecraft's life support system, which regulates his temperature and humidity by controlling the oxygen flow.

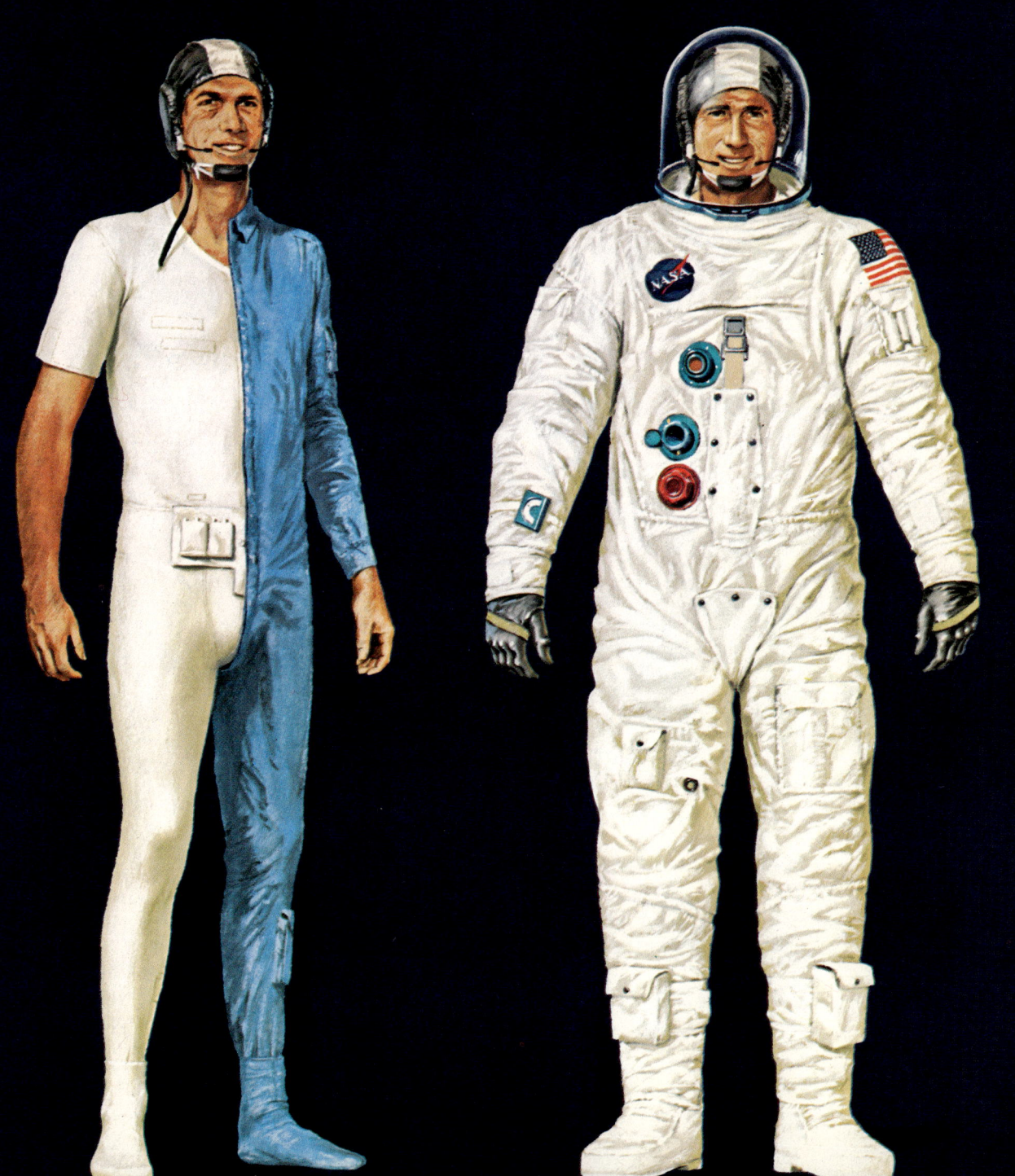

EXTRAVEHICULAR

To enter the hostile void of space, our astronaut needs a different suit, one which has better insulation and can withstand impacts from the superfast dust particles called micrometeoroids. This is his *extra-vehicular cover*, also called the *integrated thermal micro-meteoroid garment*, a long name which means that it is a combination pressure suit plus protective cover. It contains 18-20 layers of fabric, and is more difficult to flex than the similar-appearing intravehicular suit. If the astronaut intends to take only a short space walk, he can wear his EV suit over the constant-wear garment and hook up to the spacecraft's life support system, as he does with the IV suit. For moonwalking, however, he must carry his own life support system with him, the PLSS backpack and its emergency oxygen supply; these connect to a different set of fittings on the opposite side of the suit. Underneath, the constant-wear garment is replaced by the *liquid-cooling garment*, which contains a network of plastic tubes full of water circulated by a pump in the backpack. Lunar boots cover his normal space shoes, and thick lunar gloves fit over the pressure gloves. Solar glare is intense in space, so the helmet is fitted with two flip-up sun visors, as well as a cloth cover and a movable plastic screen on each side. Each lunar astronaut also carries a radio transceiver in the backpack.

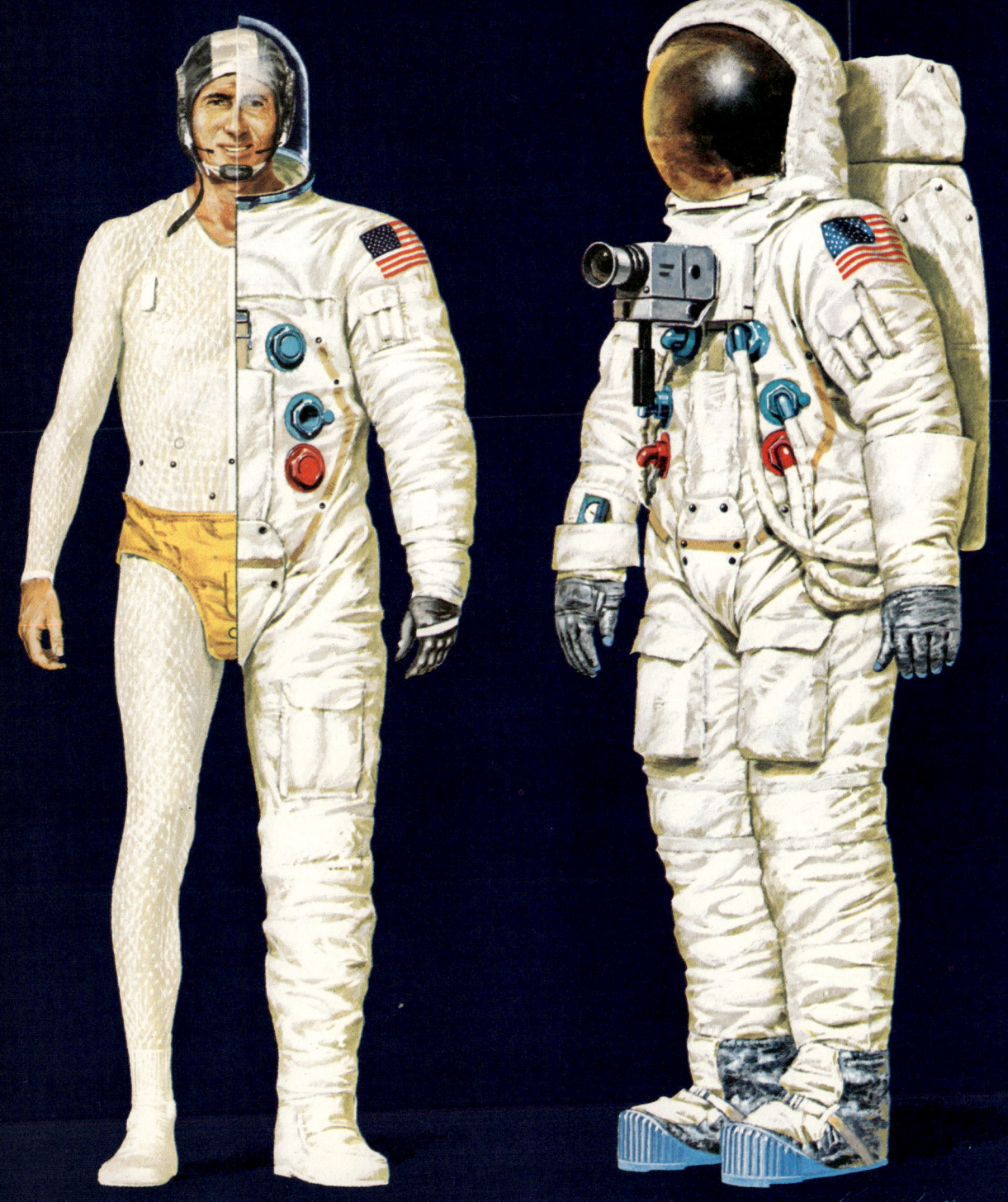

ILLUSTRATION BY KEN HODGES

APOLLO 11

This is it. Ten years of effort have been directed toward this one flight. America's massed technology—shocked into space by Russia's Sputnik—has responded with a cooperative peacetime assault, the likes of which the world has never seen. And never before in history have as many people known about a coming event. Three men will breach the unknown in an exercise as daring as the time an Italian-born commander had left Spain with three sailing ships and headed into a great void, with one purpose—to set foot upon the New World. Columbus acheived his objective. Apollo 11 will be the new spear—of unprecedented effort—yet will bear the shield of high drama. CM Pilot Michael Collins, an American, was born in Rome, Italy. His flagship is named *Columbia,* after the moon-rocket in Jules Verne's century-old fantasy. The LM *Eagle* is represented by Apollo 11's official emblem of dark blue sky behind a silver moon, over which swoops an eagle, clutching not its prey, but an olive branch. Going to the moon will be a curiosity about it as old as man's recorded history. Now, in mid-summer 1969, it is an international race: Russia's unmanned Luna 15 is launched three days earlier... America has its faith in Armstrong, Aldrin and Collins.

NEIL A. ARMSTRONG

SPACECRAFT COMMANDER/ Unassuming and dedicated since his early years as a Boy Scout, Neil Armstrong is now one of the finest pilots of his generation. He has flown the X-15 rocket aircraft more than 4000 miles an hour; yet is a superb glider pilot. As a U.S. Navy pilot, he flew 78 combat missions in Korea. He has BS and MS aeronautical engineering degrees; and was backup Command Pilot for Gemini 5, 11, and Apollo 8 missions. Armstrong showed exceptional skill, as Command Pilot, in landing the crippled Gemini 8.

MICHAEL COLLINS

COMMAND MODULE PILOT/ Mike Collins has logged more than 4200 hours flying time. He was graduated from Saint Albans School in Washington, D.C., and the U.S. Military Academy at West Point. Among his honors as an Air Force officer, Collins has received the Air Force Distinguished Flying Cross. An expert with performance, stability and control problems of jet fighters, he was backup pilot for Gemini 7 and was pilot during the record-setting rendezvous flight of Gemini 10. Collins is the astronauts' handball champion.

EDWIN E. ALDRIN, JR.

LUNAR MODULE PILOT/ "Buzz" Aldrin, like Collins, received his Air Force wings after graduation from West Point. Aldrin flew 66 combat missions over Korea, and was credited with destroying two MIG-15 aircraft. He has commanded a German-based USAF fighter wing, and has received nearly every aviation award, plus the highest awards of 10 foreign nations. Called the "best scientific mind in space" by his colleagues, Buzz has a Ph.D. from MIT in astronautics; was backup CM Pilot for Apollo 8, and the pilot for Gemini 12.

APOLLO 11 Pre-launch

Part of stringent Apollo 11 training is practice with the centrifuge at Houston's Manned Spacecraft Center. The man who will keep a lonely vigil in the Command Module *Columbia,* CM Pilot Michael Collins, sits in the centrifuge gondola (**1**) at the center's Flight Acceleration Facility in Building 29. Collins masters proficiency in another simulator, too, this one for the Command Module (**2**). The CM simulator provides Collins with actual duplication of a rendezvous and docking maneuver—another in the long line of ground simulations—and crucial for a lunar landing. Practicing in the spacesuit, as he hopes to appear on the lunar surface, is Mission Commander Neil Armstrong. Note the camera attached to his chest area (**3**), providing freedom for the arms. He also simulates a first lunar step (**4**) during another session. Note his right foot on a mockup Lunar Module foot pad. Meanwhile, the S-IVB stage is being mated. The instrumentation unit of the space vehicle contains the following systems: environmental control,

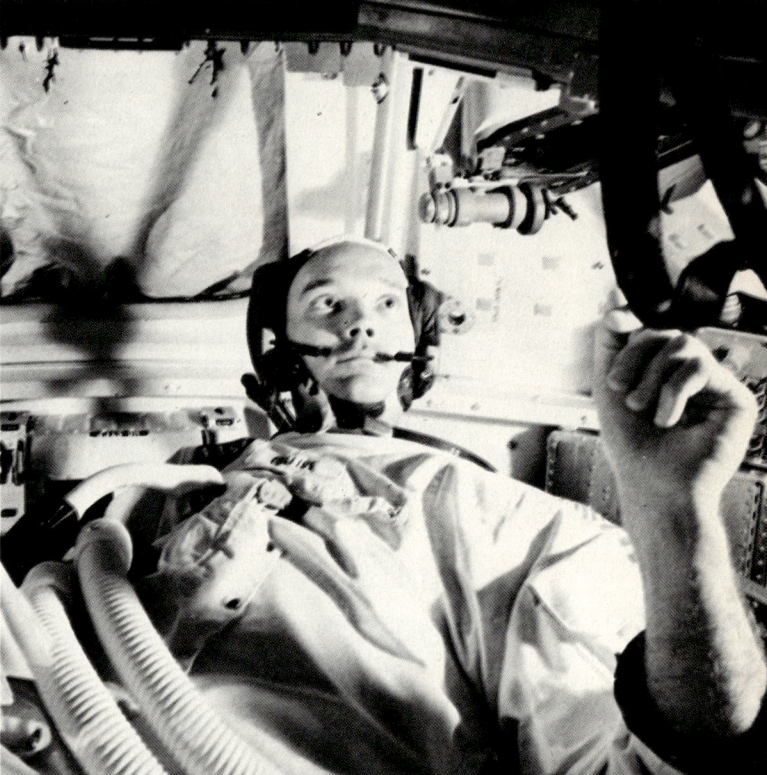

guidance, electrical and structural instrumentation; and (**5**) is now raised for mating. The fully assembled Saturn V is called Apollo/Saturn 506 (A/S 506); the Command Module *Columbia* is No.107, and the Lunar Module *Eagle* is No.5 (LM-5). Now it is time for mating the Command/Service Module to A/S 506; and the module (**6**) is lowered. Loaded weight at launch for the CSM will be more than 75,000 pounds, and for the fully loaded Saturn V launch vehicle at launch (including propellant): 5,625,000 pounds. It's hard to realize that A/S 506—shown here at the moment of accepting the CSM (**7**)—will stand, fully loaded at lift-off, some 60 feet higher than the Statue of Liberty on its pedestal, and will weigh 13 times as much. An exterior view of that height can be seen with rollout from the Vehicle Assembly Building (**8**) at Cape Kennedy. The space vehicle's massive transporter heads slowly down the 3.5-mile Crawlerway to Launch Complex 39-A. "Slowly" means at speeds not to exceed 1 mile an hour. It is 12:20 PM EDT several weeks prior to launch, and the rollout begins (**9**) which will take the transporter to the pad by 6:30 PM. The Crawlerway is 130 feet wide and 7 feet thick; composed of hydraulic fill, selected fill, graded lime rock and asphalt sealer coat; topped with 8 inches of graded river rock. But the moment has come—all of a sudden, it seems—it is nearly 6:54 AM, lift-off day; and Neil Armstrong (front), is followed by Mike Collins, as he approaches the spacecraft (**10**).

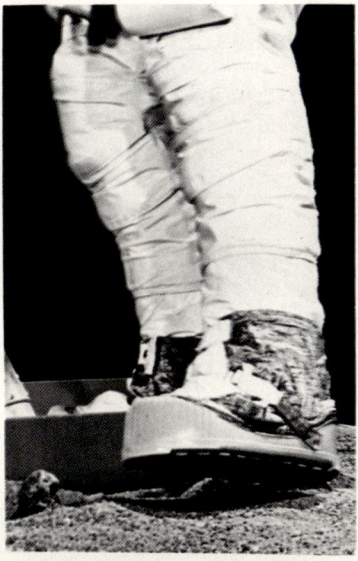

APOLLO 11
Launch

Today, July 16, 1969, there are more than 6000 guests of NASA on hand to witness the flight of Apollo 11. Also at the Cape, there are at least 1750 journalists. And on the nearby Indian and Banana rivers, there are more than 3000 boats of many descriptions, having come by sea to watch the great adventure to the moon. The crowds don't know of two minor problems that have developed in the ground equipment. But a leaky valve and a faulty signal light have been corrected. It is 9:31 AM EDT, eight and nine-tenths seconds before launch time. A roar comes from the launch pad. One by one, the five Saturn first stage engines belch fire. A witness on the VIP section of the reviewing stand is instantly aware of perspiration beads on upraised foreheads, as a summer sun boils across the Florida landscape. The Apollo access arm has been retracted, and for two seconds the space vehicle builds up thrust. Now the hold-down clamps release, and the space vehicle rises from the pad—slowly at first. Watches read

9:32 AM. The launch is only 724 milliseconds late. But most witnesses don't know that. They hear only the booming waves of thunder, as the space vehicle passes the top of the umbilical tower. Former President Lyndon B. Johnson, without sunglasses, squints against the fire (**1**) and watches the first lunar landing mission—on its way at last. On his right is Mrs. Johnson, while Vice-President Spiro Agnew, wearing dark glasses, is on Mr. Johnson's left. Other distinguished persons are there, too. The list includes more than 800 foreign journalists, with U.S. officials numbering some 205 Congressmen, 30 Senators, 19 Governors, at least 50 Mayors and nearly 70 Ambassadors. As Apollo 11 hurtles to the heavens, a trance seems to envelope the press section. There (**2**), a photo is taken with a Hycon KA-85A panoramic camera, using 70mm Ektachrome MS film. Exposure is 1/1000 second at f5.6. At that moment, exposures of a different kind are taking place inside the spacecraft. Buzz Aldrin had once remarked that men could remember not to throw a particular switch, but that others theoretically could be thrown at the wrong time. The Command Module panel display includes 24 instruments, 566 switches, 40 indicators and 71 lights. But the lift-off has been faultless; especially within the Launch Control Center (**3**). A close-up at the moment of lift-off (following page 130) shows the Saturn V's awesome power, while an airborne photo, 2½ minutes after lift-off, sees the space vehicle at an altitude of some 38 miles (page 131).

APOLLO 11
In-flight

Reportedly, Russia's Luna 15 is approaching lunar orbit. Apollo 11, however, with all systems "Go," is now hurtling into space at 17,358 feet per second—nearly 12,000 miles per hour. Altitude is 95 miles. Ten minutes away from Translunar Injection (TLI), the view below is of a distant, dying storm over the Pacific Ocean, west of Australia. The clouds in the vast panorama look like layers of spun wool, as darkness creeps to the horizon, and only a spot still mirrors the sun's reflection (**1**). Translunar Injection is the big burn of the Saturn, designed to send Apollo 11's crew on a long moon-coast, following engine shutdown, so that some 200,000 miles "down" to the moon will not require engine propellant. And, there is only a 4-hour margin in which to make a correct TLI. During the burn, the spacecraft's velocity leaps to 35,575 feet per second. Shortly after 1 PM EDT, on the fourth day following launch, the spacecraft passes completely behind the moon, losing radio contact with earth for the first time. After the first revolu-

tion around the moon, the astronauts emerge from the dark side to see the rising earth (**2**) just above the lunar horizon, its circular cloud patterns looking like knurled whipped cream, 240,000 miles away. The astronauts are impressed with this view of the earth, in contrast to the oblique view of the lunar far side (**3**) which they have just seen. The large central crater is International Astronomical Union No.308, with a diameter of about 50 miles. Nearly three hours later, a 35-minute telecast of the moon's surface begins, as the spacecraft passes westward along the eastern edge of the moon's visible side. The next day, July 20, Aldrin crawls into the Lunar Module and powers-up the engine. At 1:46 PM the LM is separated from the Command Module, CM pilot Collins firing rockets for initial release (**4**). Collins watches the Lunar Module intently (**5**) as it drifts away in the vacuum of space—becoming probably the loneliest man in the universe as he later passes the far side of the moon. The undocking—the crucial moment—follows the 12th orbit of the moon, at about 100 hours and 40 minutes, ground elapsed time (GET). Collins reports from *Columbia* that ''Everything's going just swimmingly. Beautiful!'' An hour later, Armstrong and Aldrin, flying in the LM, feet first and face down, fire the descent engine for the first time. Mission Commander Armstrong reports, ''The *Eagle* has wings.'' The LM enters its descent orbit insertion. The race between Luna 15 and Apollo 11 reaches classic proportions—two hours from landing.

APOLLO 11 Exploration

Speculation is now that Luna 15 has left lunar orbit and will attempt to land; perhaps retrieve a quantity of lunar soil and return to earth before Apollo 11. *Eagle's* powered descent to the lunar surface begins at 50,000 feet above the moon. On earth, discussions between NASA and the Soviet Institute of Science have by now established that Luna's orbit will not conflict, but about a landing attempt the Soviets will say nothing. On earth, approximately 30 million American households have their TV sets tuned to the drama in outer space; with a similar audience worldwide numbering some 529 million. At 668 feet above the moon, *Eagle* is dropping at 22 feet per second. Inside *Eagle*, all systems are working perfectly. But then it happens...something goes wrong. The landing computer is overloaded with jobs to do. It refuses to work any more landing equations. A machine has failed... and man must now take over, for guidance is critical... *Eagle* has been programmed inadvertently to land in a 600-foot

wide crater, strewn with boulders the size of automobiles. Mission Commander Neil Armstrong (his heartbeat rising from a normal 77 to 156) seizes manual control of the LM. Buzz Aldrin gives him altitude readings: "...Four hundred feet, down at nine... Got the shadow out there...75 feet, things looking good...Lights on...picking up some dust...Faint shadow...Four forward. Four forward, drifting to the right a little. Contact light. Okay, engine stop." *Eagle* touches with firm impact, in a one-sixth gravity free-fall from about 5 feet. It is 4:18 PM EDT, July 20, 1969. There is silence in Houston—then over the monitors, a word from far away (**1**): "Houston. Tranquility Base here. The *Eagle* has landed." The Capcom responds: "Roger, Tranquility. We copy you on the ground. You've got a bunch of guys about to turn blue. We're breathing again. Thanks a lot." Cheering and tears are spontaneous. The next day, the astronauts venture onto the moon's surface. At 9:56 PM, a TV transmission shows the world Armstrong's first step (**2**) and hears his words: "That's one small step for a man, one giant leap for mankind." (**3**) Only 20 minutes behind, Aldrin descends to the moon (**4**) and stands next to the flag (**5**), before deploying the (**6**) Modularized Equipment Stowage Assembly (MESA). The first thing to touch the moon, but by now a squashed Lunar Sensing Probe (**7**), is visible under the LM's footpad. Also visible in the visor of Aldrin's helmet is the photographer, a tiny speck (**8**) named Neil Armstrong.

APOLLO 11
Exploration

The surface at Tranquility Base is stark, yet smooth. There are small rocks of every description, and the dull-gray soil has the top-surface consistency of talcum powder—leaving clearly etched bootprints, and clinging to the soles of the boots. Underneath, however, the soil is extremely hard-packed. Armstrong thinks that the small, rounded craters resemble the swells of an ocean (**1**); remarking that the surface obscures the horizon. A view to a man places the horizon less than 2 miles away, and from inside the LM, about 4 miles distant. But at least *Eagle* is on the moon... Russia's Luna 15 has now crashed in Mare Crisium (The Sea of Crises), about 500 miles away from *Eagle*. It is clearly a U.S. triumph. Now, having reached the moon, science is more important. The Laser Ranging Retroflector, sounding to the watching world like an artifact of science fiction, is carried in Aldrin's gloved hand (**2**) to a suitable spot. The reflector is set-up (**3**) and will allow for precise measurement of the earth-moon distance. Next,

Aldrin finds a good place for the Early Apollo Scientific Experiments Package (EASEP), not too far (**4**) from the LM. As Aldrin attends to the solar wind experiment (**5**), it appears evident that the lunar landscape has great contrasts of light and dark. The solar wind experiment (**6**) measures minute particles which bombard the moon constantly from outer space. A major experiment is to deploy the Passive Seismic Experiment (PSEP), which will detect and record moonquakes, meteorite impacts or volcanic eruption. The PSEP completes the first scientific station, permanently installed on the moon; and Aldrin (**7**) now heads back to the LM. Aldrin's bootprint penetrates lunar soil a quarter- to a half-inch, giving a firm response, but causing a slight sideways motion that invites care. It is the care of watching one's balance in the one-sixth earth gravity. Time passes all too quickly, as Aldrin discovers that he has completed 2 hours and 15 minutes on the lunar surface; and Armstrong, 2 hours and 35 minutes.

Lift-off is approaching; the crucial ascent to lunar orbit rendezvous with *Columbia*—and the trip home. When the time comes for *Eagle* to spread its wings once more, the astronauts will leave behind the permanent scientific research station; and a plaque on one leg of the descent stage, reading: "Here men from the planet earth first set foot upon the moon—July 1969 A.D.—We came in Peace for all Mankind." Also left will be a 1½-inch silicon disc (**8**) bearing goodwill messages from many heads of state.

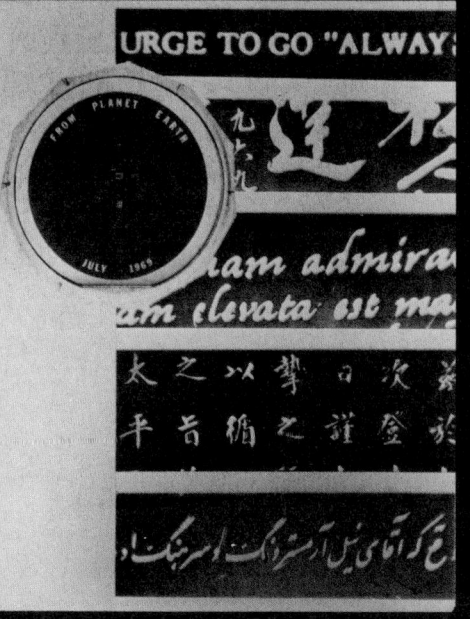

APOLLO 11
In-flight

At 1:11 AM EDT, July 21, the astronauts close the hatch, and begin removing their Portable Life Support Systems. At 4:25 AM they are told by the Capcom to go to sleep, after working on final housekeeping details and answering a number of questions about the geology of the moon. As the first men on the moon sleep, CSM pilot Michael Collins circles overhead in his orbit. Shortly before *Eagle's* astronauts are aroused by Mission Control, it is observed of Collins: "Not since Adam has any human known such solitude as Mike Collins is experiencing during this 47 minutes of each lunar revolution when he's behind the moon with no one to talk to except his tape recorder aboard *Columbia.*" Then the moment of lunar lift-off is at hand. At 1:54 PM the ascent engine is started. *Eagle*, using its descent stage as a launch pad, begins rising, and reaches a vertical speed of 80 feet per second at 1,000 feet altitude. Brought along are soil samples, and the aluminum foil with the solar wind particles it has collected. By leaving

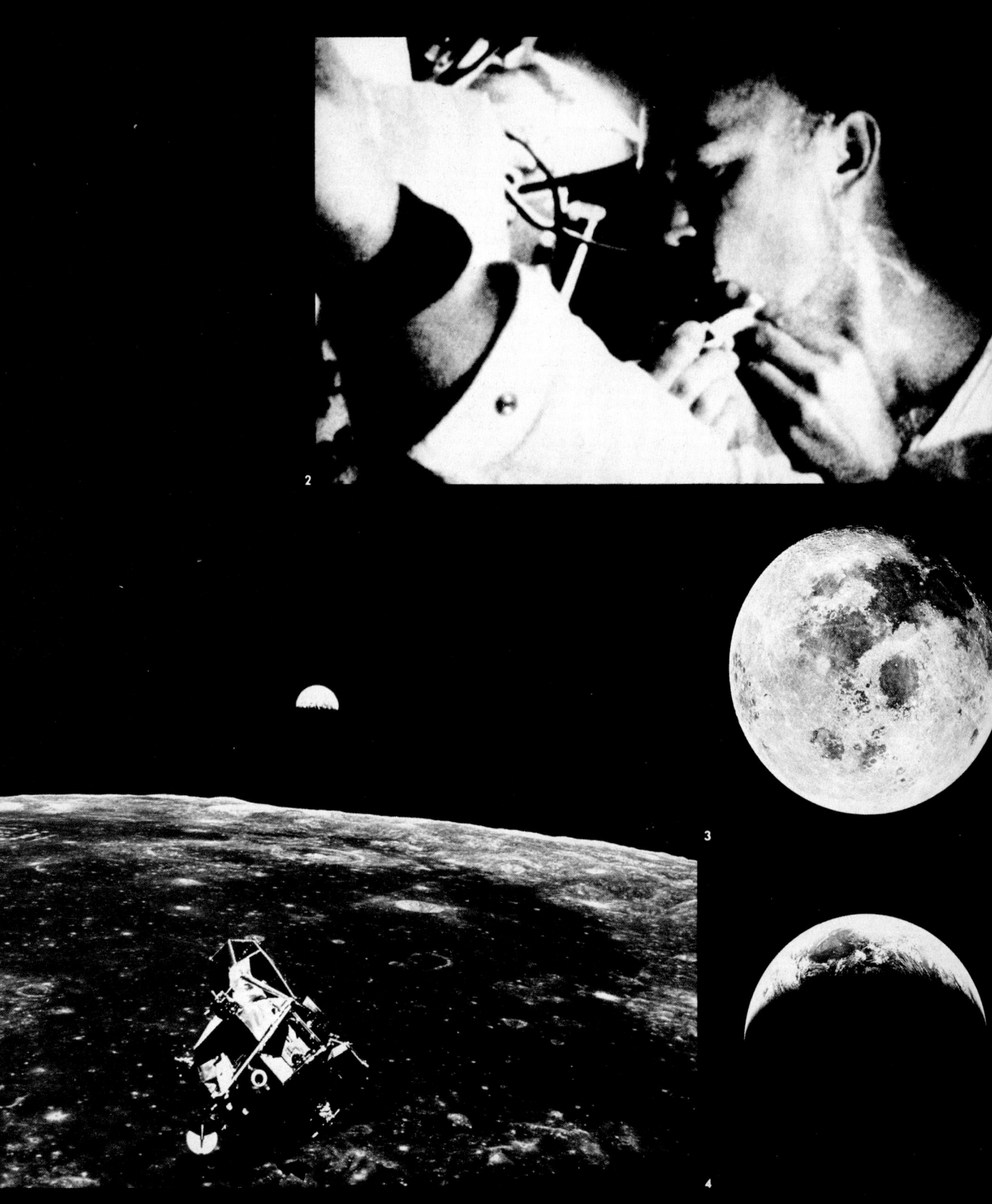

ing the descent stage, the LM is some 5,076 pounds lighter (**1**) as it heads for redocking with the Command Module. It is a new sight for man—the oasis earth from a returning spacecraft, having touched the moon. It is something not done before—as was the first telephone call from an American President via the Capcom to men on the moon. But now, in the vastness of space, there is precious little time to reflect. The work that needs to be done takes first priority over any tendency to dwell on *Eagle's* success. Once man had dreamed about the moon—as a thing of poem and legend. The conquest of the moon took nearly two million years of man's development. And though man had gotten there at last . . . it is still an almost unthinkable thought; a dream, almost, that disappears in the even strokes of Mike Collins' razor (**2**). After all, *Eagle* and *Columbia* have docked successfully, and the moon is 10,000 miles behind (**3**). The "aliens" are going back to their earth (**4**), now a brilliant ornament behind the moon's shadow. At 7:03 PM on July 23 some 90,000 miles from earth, Neil Armstrong smiles during the final color TV transmission from space (**5**). The astronauts are extremely relaxed, as Aldrin shows a gyroscope principle with a can of food (**6**) in zero gravity. Now the CM is hurtling some 4300 feet per second through the vacuum of space, and Aldrin shows how to make a sandwich under the same zero-gravity conditions, obviously enjoying it (**7**)

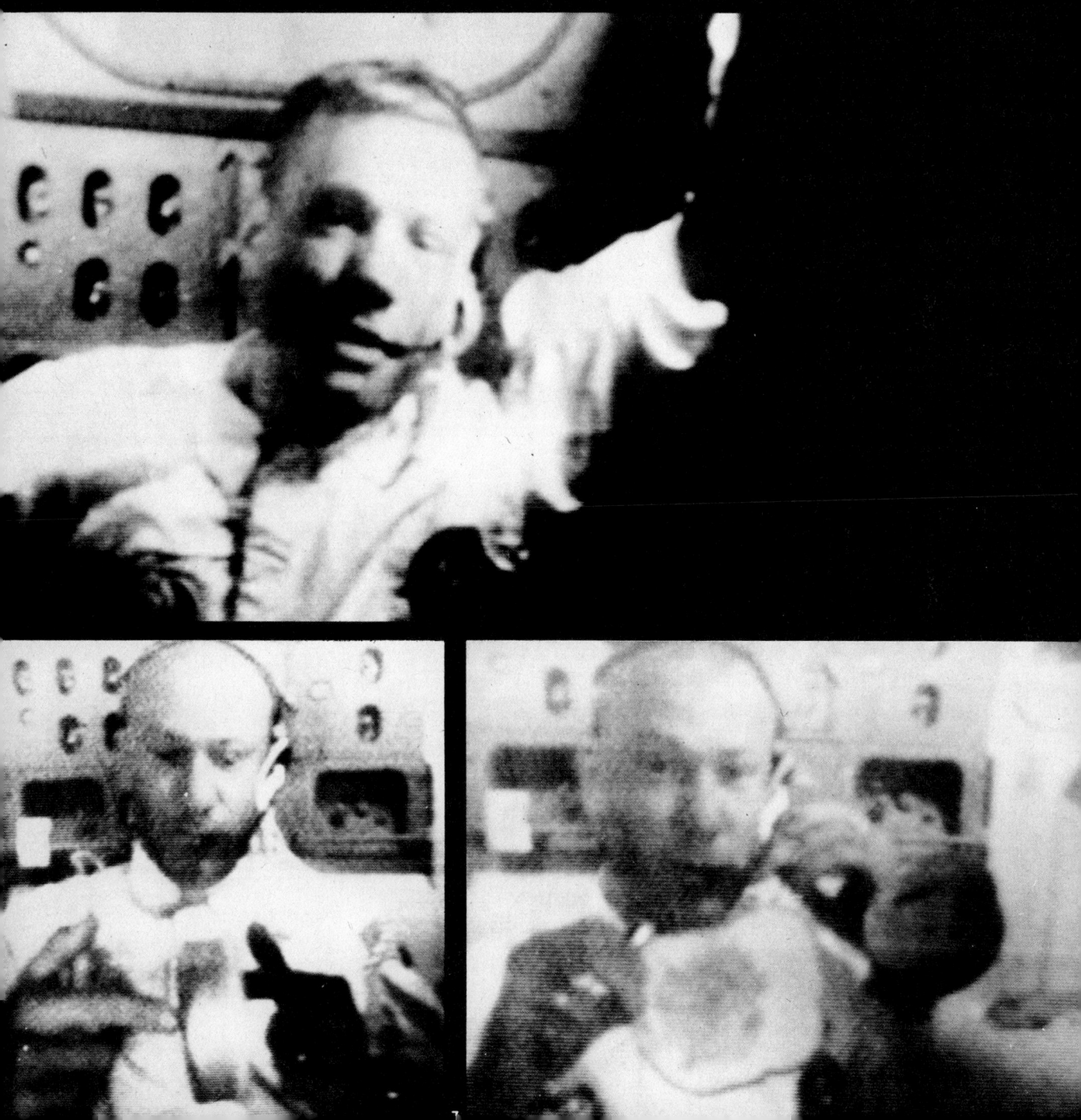

APOLLO 11
Splashdown

The next day, splashdown day, the crew of Apollo 11 awakens at 6:47 AM. At 21 minutes after 12 noon, the Command and Service Modules are separated and 14 minutes later the Command Module re-enters the earth's atmosphere. It is now 12:51 PM, July 24, and a strange thing coming down through the clouds is recognized by the television viewers of the world as *Columbia*—no longer a thing from a Jules Verne fantasy—but a capsule, the splashdown and emotional reverberation of which can be felt worldwide through the instant magic of the electronic media. *Columbia* hits the water beneath its parachutes some 825 miles southwest of Honolulu and about 13 miles from the recovery ship, U.S.S. *Hornet*. Scarcely more than 30 minutes later, the astronauts emerge from the spacecraft in isolation suits and are sprayed with disinfectant to guard against the possibility of their contaminating the earth with moon "germs." At 1:57 PM, the recovery helicopter arrives on the *Hornet's* elevator, and a technician

holds the Mobile Quarantine Facility door for the astronauts (**1**). This moment, as seen from the side (**2**), shows (left to right) Aldrin, Armstrong and Collins. Once inside their Mobile Quarantine Facility aboard the *Hornet*, the three (from left, Armstrong, Collins and Aldrin) share a joke with well-wishers (**3**). One of their ardent fans is President Richard Nixon (**4**), who happily shares light banter with the world's space heroes. Nixon tells them seconds later: "This is the greatest week in the history of the world since the Creation.... As a result of what you have done, the world's never been closer together... We can reach for the stars just as you have reached so far for the stars." The Quarantine Trailer is a remarkable concept, sealing the astronauts inside a cocoon until their arrival at the Lunar Receiving Laboratory, Houston, on July 27. But first, the *Hornet* steams into Pearl Harbor for transfer of the trailer to a C-141 Jet transport. Off-loading the trailer at Pearl is a precise operation (**5**) that naturally has its share of public curosity (**6**). Inside the Mobile Quarantine Facility (**7**), astronauts (from left) Collins, Aldrin and Armstrong relax and learn what the newspapers have said about their moon voyage, as the C-141 prepares to swallow its precious cargo (**8**). Perhaps the astronauts are wondering, in a flashback of thought, how the splashdown moment must have looked (**9**) at Mission Control. After pandemonium had subsided, it was time for cigars (**10**) by NASA and Manned Spacecraft Center officials. "Apollo 11 has done it!"

APOLLO 11
Splashdown

Another flashback includes the precious recovery and stowage of the Command Module, now bathed in the shadows of *Hornet's* hangar deck (**1**); and a view of the U.S. Marines standing guard (**2**). Behind them is the scorched skin, having been subjected to a screaming plunge through the earth's atmosphere at a speed approaching 25,000 miles per hour. Note the flotation bags still attached to the scarred shell of *Columbia*. As a concept of space transport, the CM has proved its worth by carrying three human beings more than a million miles since July 16. And then there are the wives; the emotional impact felt and carried on their shoulders for the men they love. Nothing can document something so personal. Each could speak now to her husband, even though they are protected and out of finger-tip touch behind the isolating glass of the Mobile Quarantine Facility. Janet Armstrong and the couple's two sons had watched the lift-off from a sailboat on the Banana River. Pat Collins and Joan Aldrin watched from

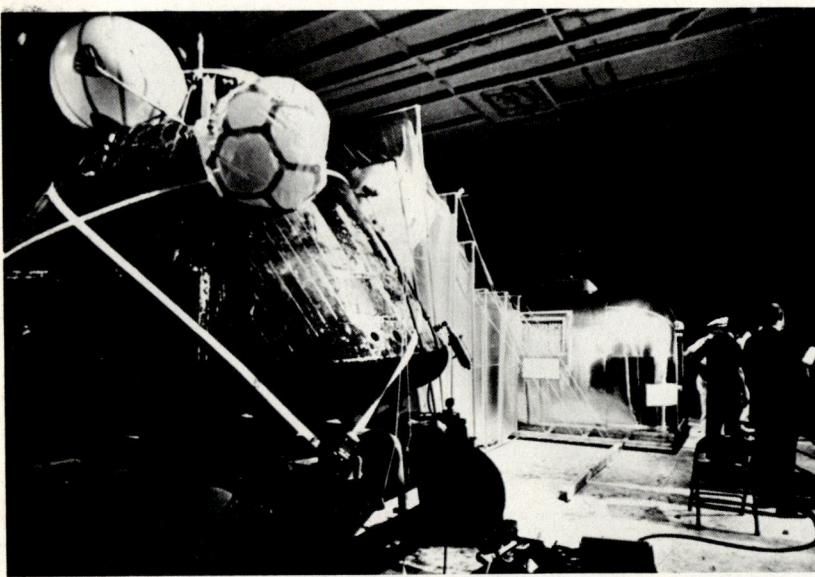

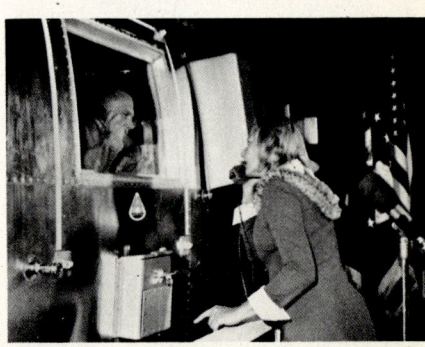

Houston TV. Now, on July 27, the three wives could meet their husbands face-to-face, as the C-141 Starlifter lands at Ellington Air Force Base in Houston. It is after midnight, but Joan Aldrin isn't tired. There is relief and inexplicable joy in her face and in her words. With a patched-in telephone to the Mobile Quarantine Facility (**3**), Joan and Buzz share a very private moment. Man's first mission to the moon is a precise acheivement and a wordly triumph. It has lasted 195 hours, 18 minutes and 35 seconds, or about eight days. Of all the space missions, Apollo 11 has been the most trouble-free mission so far; almost totally on schedule and successful in every way. To those who have raised the question of cost, it can be pointed out that many of the tallest cathedrals in Europe were built in times and ages of surrounding poverty. And though poverty is an eternal obstacle of mankind, so too is that precious something in man's soul . . . a something that believes in building monuments to his capacity to dream. The success of Apollo 11— to be able literally to walk on the moon–is a great moment for his species. These are some of the thoughts behind the cohesive worldly acclaim—even in the country of the ill-starred Luna 15—and especially behind the standing ovation given to Armstrong, Aldrin and Collins by the Joint Houses of Congress. The ovation has followed shortly after Collins' remarks (**4**); and is mirrored in a ticker tape parade (**5**), billed by the City of New York as the largest in its history: a heroes' welcome.

APOLLO 11
Post-flight

With the splashdown of the Command Module *Columbia*, the world learns not of an end, but of a new beginning. At the Manned Spacecraft Center's Lunar Receiving Laboratory, one of the moon rock samples (**1**) collected during the more than two-hour lunar walk, is now receiving its analytical baptism. The texture appears to have cooled quickly near the lunar surface—a premise readily accepted by petrologists. But they are not sure whether melting has resulted from volcanic activity or from meteoric impact. It also appears that this ¾-inch-wide rock sample contains identifiable earth elements: iron, silicon, aluminum, oxygen, magnesium, even titanium. But the proportions and chemical bonds are different from anything on earth. Also busy are NASA botanists, who test lunar soil for its ability to sustain growth—as with these sorghum and tobacco plants (**2**) in the MSC's Plant Laboratory. The flight of Apollo 11 has signaled the end of an era. The moon is no longer the same, for man has touched it at last.

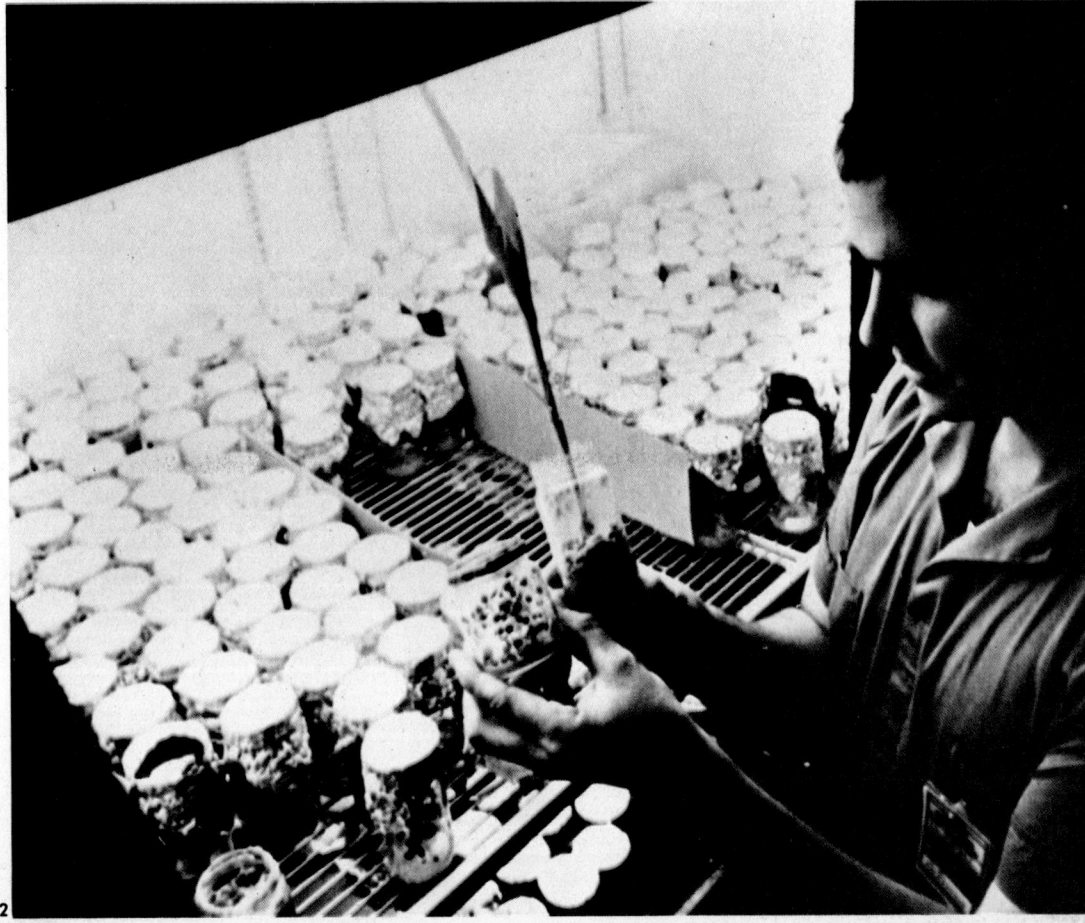